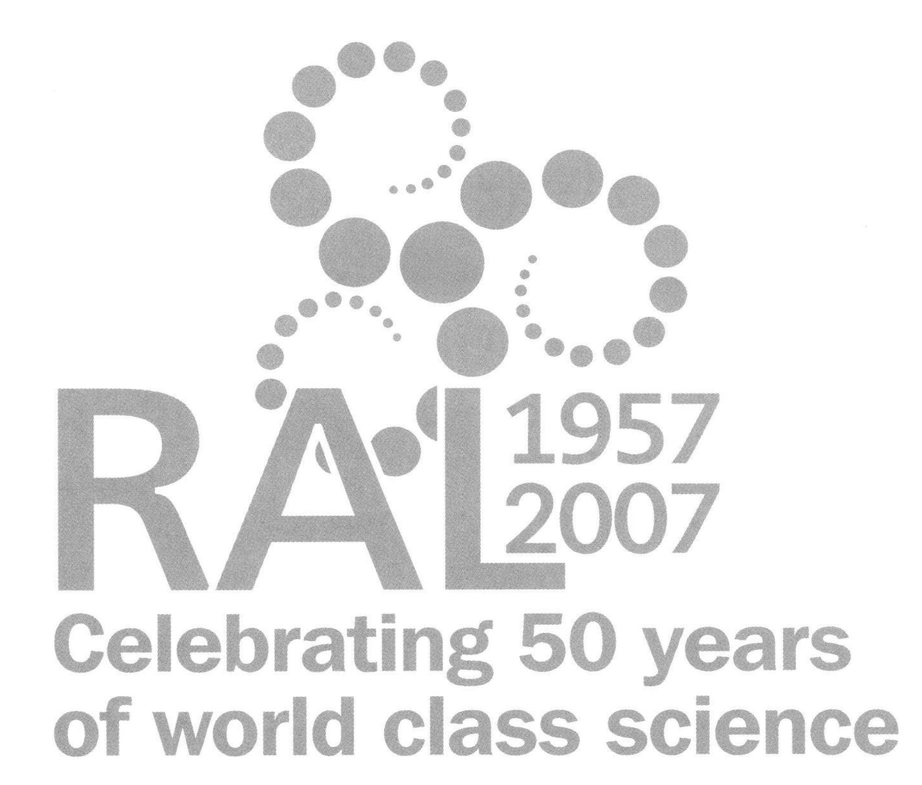
RAL 1957 2007
Celebrating 50 years
of world class science

Fifty years of excellence

From its origins 50 years ago, the Rutherford Laboratory has been renowned for its excellence in supporting UK Science.

The National Institute for Research in Nuclear Science was formed in 1957 to operate the Rutherford High Energy Laboratory under the Directorship of Dr Gerry Pickavance. The 50 MeV proton linear accelerator – the PLA – was transferred from the UKAEA to the new laboratory which, under Dr Godfrey Stafford's guidance, set about designing a national machine for UK particle physics, complementing the accelerators being developed at Brookhaven in the USA and CERN. Nimrod – the mighty hunter – was commissioned in 1963 and operated for the following 15 years, successfully developing the capabilities of the UK particle physics community. Rutherford Laboratory, through its work on bubble-chambers, electronic detectors and instrumentation, gained a world wide reputation for excellence that it has to this day.

Stafford succeeded Pickavance as Director in 1969. A decade later, as Director General, he oversaw the growth of the laboratory in two ways: through the amalgamation of the Atlas Laboratory – the centre for academic computing on the Chilton site – and the merger with Appleton Laboratory, which had a distinguished history in radiopropagation and space research, to create the Rutherford Appleton Laboratory (RAL).

But it was Dr Stafford's expansion into new areas that is his lasting legacy: the development of neutron beams for research in the physical and life sciences, and the creation of the Central Laser Facility.

Rutherford Appleton Laboratory 2006.

In 1971, a Neutron Beam Research Unit was established at the laboratory to support the burgeoning community of university scientists who used neutron beam facilities for their research. Its first task was to develop the case for a high flux reactor to be built on the Chilton site. In the event the reactor project did not proceed, but the Unit proved to be just the right grouping to catalyse the development of a new approach to the provision of neutron beams. Thus the Spallation Neutron Source project – later named ISIS by Prime Minister Margaret Thatcher in 1985 – was born. Under the inspiring leadership of Geoff Manning, the Laboratory's expertise in accelerators, targets, detectors, electronics and computing was re-directed to construct a world leading facility for neutron science, redefining the future of neutron scattering world wide. In 1981, Manning succeeded Stafford as Director.

The laboratory also became the home of a high-power laser facility for UK academic plasma physics research. A judicious choice of parameters, and excellent partnerships between Rutherford staff and the academic community, made Vulcan a highly competitive facility on the world scale. The development of chirped pulse amplification, and subsequent power upgrades to petawatt capabilities have established RAL as a world leading centre for laser research.

Dr Paul Williams completed the transformation from the laboratory's single mission origins, to the diverse multi-disciplinary international centre of excellence that it is today. As Chairman and Chief Executive, he amalgamated RAL with its sister laboratory in Daresbury to create the Council for the Central Laboratory of the Research Councils.

The relevance of the laboratory's programme to the needs of industry and the wider society grew in importance under the stewardship of Dr Bert Westwood and Dr Gordon Walker. In recent years, Professor John Wood has been pivotal in establishing the laboratory in the context of the Harwell Science and Innovation Campus, emphasising its potential in delivering trained manpower and economic impact in addition to world-class science.

The 50th anniversary sees the laboratory as vibrant and as relevant as ever, a source of support to UK academia and industry, home to a diverse range of technologies with exciting futures in condensed matter science, lasers and space science, with the possibility of hosting major international facilities for neutrons and for neutrinos. It continues to develop cutting edge technologies which underpin its internal programmes and offer solutions for society.

After 50 years of excellence, a further golden age for the laboratory beckons on the horizon, as a jewel in the crown of UK science and the linchpin of the Harwell Science and Innovation Campus.

Andrew Taylor
Rutherford Appleton Laboratory

November 2007

Laboratory Directors

1957-1969	Dr Gerry Pickavance
1969-1979	Dr Godfrey Stafford
1979-1981	Dr Godfrey Stafford
1981-1986	Dr Geoff Manning
1986-1994	Dr Paul Williams
1994-1995	Dr Paul Williams
1995-1998	Dr Paul Williams
1998-2000	Dr Bert Westwood
2000-2001	Dr Gordon Walker
2001-2007	Professor John Wood

Dr Gerry Pickavance

Dr Godfrey Stafford

Dr Geoff Manning

Dr Paul Williams

Dr Bert Westwood

Dr Gordon Walker

Professor John Wood

Particle Physics

Simulated event of the collision of two protons in the ATLAS Experiment CERN, viewed along the beam pipe. The colours of the tracks emanating from the centre show the different types of particles emerging from the collision.
Credit: CERN

The Rutherford High Energy Laboratory was established in 1957 to build a 7 GeV proton synchrotron for research in particle physics. The machine came on line in 1963 and the many experiments carried out until its closure in 1978 contributed richly to the understanding and classification of the spectrum of particle states, and hence to the emergence of the now-standard view that the particles we see are not 'elementary' at all, but made of quarks. From its outset Nimrod was planned as a facility for the UK university community, who made extensive use of it, and the founding Director, Dr Gerry Pickavance, paid much attention to this aspect of the laboratory's work and life. He emphasised the importance of "a good scientific atmosphere in the national laboratory itself, which should not be a mere service station."

On its closure Nimrod itself was converted into the spallation neutron source, ISIS, which opened another chapter in the story of the laboratory. The focus of experimental particle physics moved to higher energy centres overseas, primarily CERN. Since the mid-1970s particle physics at RAL has focussed on collaboration with university groups, ensuring that the powerful technical and engineering skills and expertise at RAL are deployed to optimal effect in the design, constructions and operation of the experiments. The many experiments to which RAL scientists and engineers have made important contributions include the UA1 experiment at CERN in the 1980s, which was a breakthrough experiment for European particle physics, and our fundamental understanding of physics. RAL has also played a prominent role in the electron-positron collider experiments at CERN and at SLAC (California), in both electron-proton experiments at DESY, and most recently in the Large Hadron Collider (LHC) experiments which will start taking data next year at CERN.

Dr Pickavance's original vision of a laboratory that underpins and enhances the UK role in particle physics continues to be fulfilled today, not just through the experiments mentioned above, but also through an increasing role in the research and development of possible future accelerators. It is conceivable that the 75th anniversary will be able to reflect on the science produced by a 'neutrino factory' sited at RAL.

Sir John Cockcroft cutting the first sod at the Chilton site, 1957.

Digging the foundations for the Nimrod synchrotron, 1957. Excavation work proceeded in a number of carefully planned phases, mindful of the fact that the Chilton site was formerly an RAF airfield. Excavations deeper than the old airfield level were undertaken with great care due to the possibilities of locating discarded munitions and other scrapped materials. A 7 GeV proton synchrotron, Nimrod was the first accelerator constructed at the Rutherford High Energy Laboratory. Nimrod was used in studies of nuclear and sub-nuclear phenomena through the creation of nuclear particles at high energies.

Nimrod synchrotron under construction, 1958.

Interior of the Nimrod synchrotron hall under construction, 1958.

516
5
Z.333

The Chilton site starts to take shape, 1958. By 1959, the Rutherford High Energy Laboratory was also home to a 50 MeV proton linear accelerator.

Cleaning of the interior of Nimrod tanks not long after the synchrotron opened, 1963. Nimrod was now the second most powerful proton synchrotron in Europe and the first in Britain to be capable of producing kaons, hyperons and antiprotons. Its cost, including buildings, was just under £11 million.

Lord Hailsham (right) and Dr Gerry Pickavance at the opening of the Rutherford High Energy Laboratory as an establishment of the National Institute for Research in Nuclear Science, 1957. By 1964, the laboratory was home to divisions for applied physics, engineering and administration, and a separate group working on a novel 20 MeV electrostatic-generator project for Oxford University. It had a permanent staff of about 1000, and a budget (for 1963-64) of about £7 million.

11
12
1
10
2
9
3
8
4
7
6
5

Alternator for Nimrod in R4, 1971. Alternators were used to power fly wheels in order to produce pulsed power on Nimrod.

Neutron counter for a Nimrod experiment, 1971.

The Cockcroft-Walton pre-injector for Nimrod, 1974. This was later used on ISIS.

Hydrogen Bubble Chamber, 1971. The bubble chamber was constructed through a collaborative effort between Imperial College, London, Birmingham University, Liverpool University and Rutherford Laboratory. Filled with superheated transparent liquid hydrogen, the bubble chamber was used to detect electrically charged particles moving through it.

Film analysis laboratory, R1, 1970. Approximately 200 people (mainly women) were employed in the film analysis laboratory. Staff analysed tracks for the Rutherford bubble chamber and for similar experiments world wide.

Computer controlled graphics display used to scan data from bubble chamber experiments in Atlas, 1970.

Scanning table in Atlas, 1968.

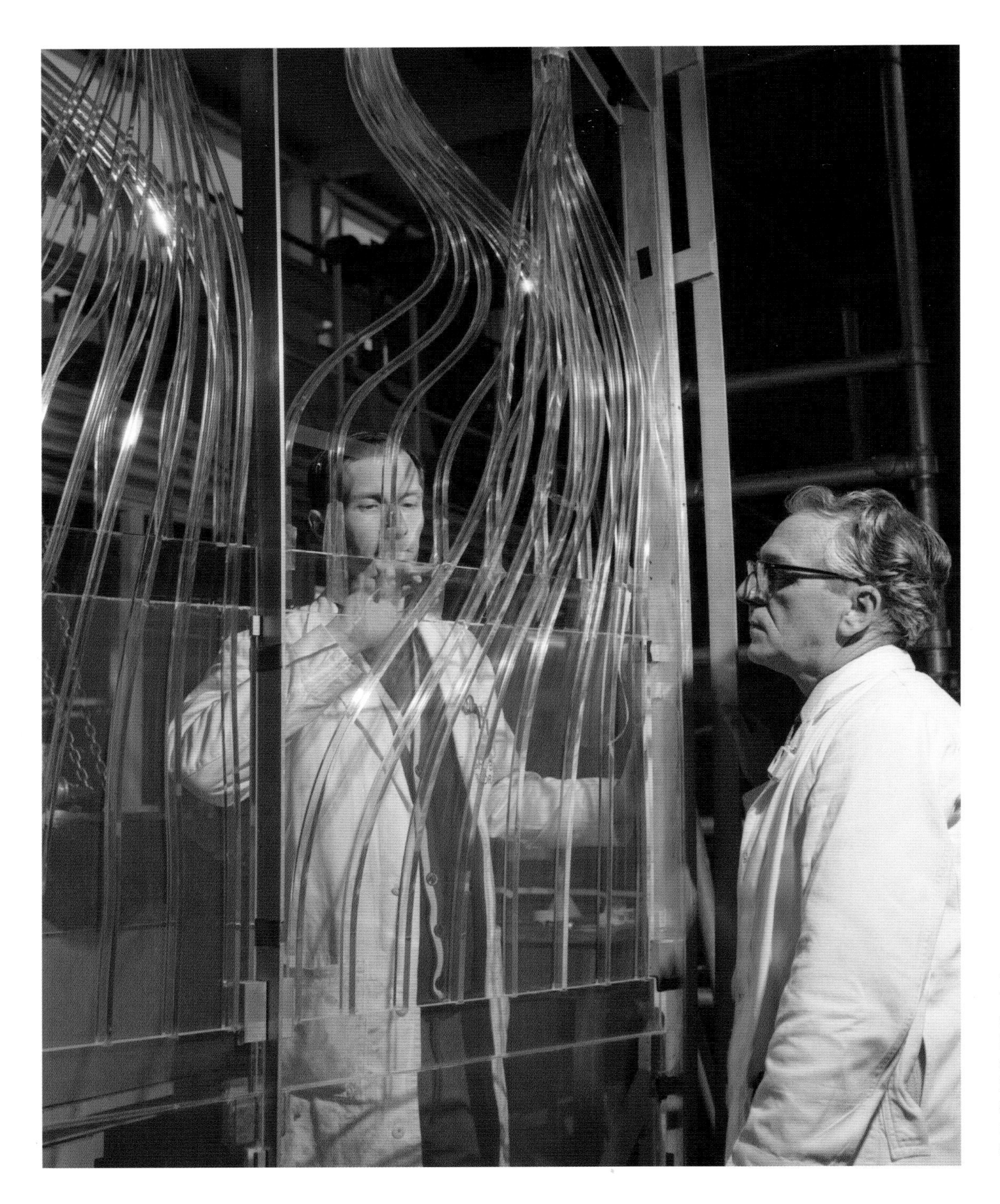

Scintillators in R12, 1971. These charged particle detectors were built at Rutherford Laboratory for use in Nimrod as well as the accelerator laboratory in Germany (DESY) and the European Centere for Nuclear Research (CERN) in Switzerland.

Main Gate, 1969.

Visit by Margaret Thatcher MP, Minister of State for Education and Science, 1972.

Library opening with Dr Gerry Pickavance, 1974.

FOR X3
LOYAL
AND
SERVICE
1978

Dr Godfrey Stafford and Dr Gerry Pickavance (behind) closing Nimrod, 1978.

Visit by Shirley Williams MP, Secretary of State for Education and Science, 1979. The Rutherford Appleton Laboratory (RAL) was formed during this year upon the merger of Appleton Laboratory (previously near Slough) with the Rutherford Laboratory.

The world's largest superconducting solenoid on its way to CERN, Geneva, 1987. The operation of the Large Electron-Positron collider (LEP) collider at CERN started in August 1989. With a circumference of 27 km, LEP is the largest accelerator yet built. LEP stopped in November 2000, but the analysis of data is still going on, with the possibility of discovering new physics phenomena.

One of the detectors, Delphi was an advanced detector. As well as having high precision and 'granularity', it has the specific ability, using the Ring Imaging Cherenkov technique, to differentiate between all the various secondary charged particles. Delphi also had an advanced silicon detector providing very precise tracking, principally in order to detect very short lived particles by extrapolating the tracks back towards the interaction point. Design and construction of the Delphi detector took 7 years, much of which was done by the Rutherford Appleton Laboratory's particle physicists.

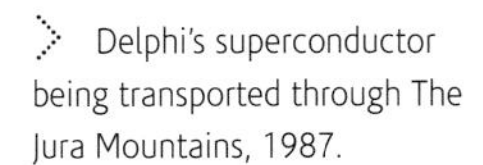
Delphi's superconductor being transported through The Jura Mountains, 1987.

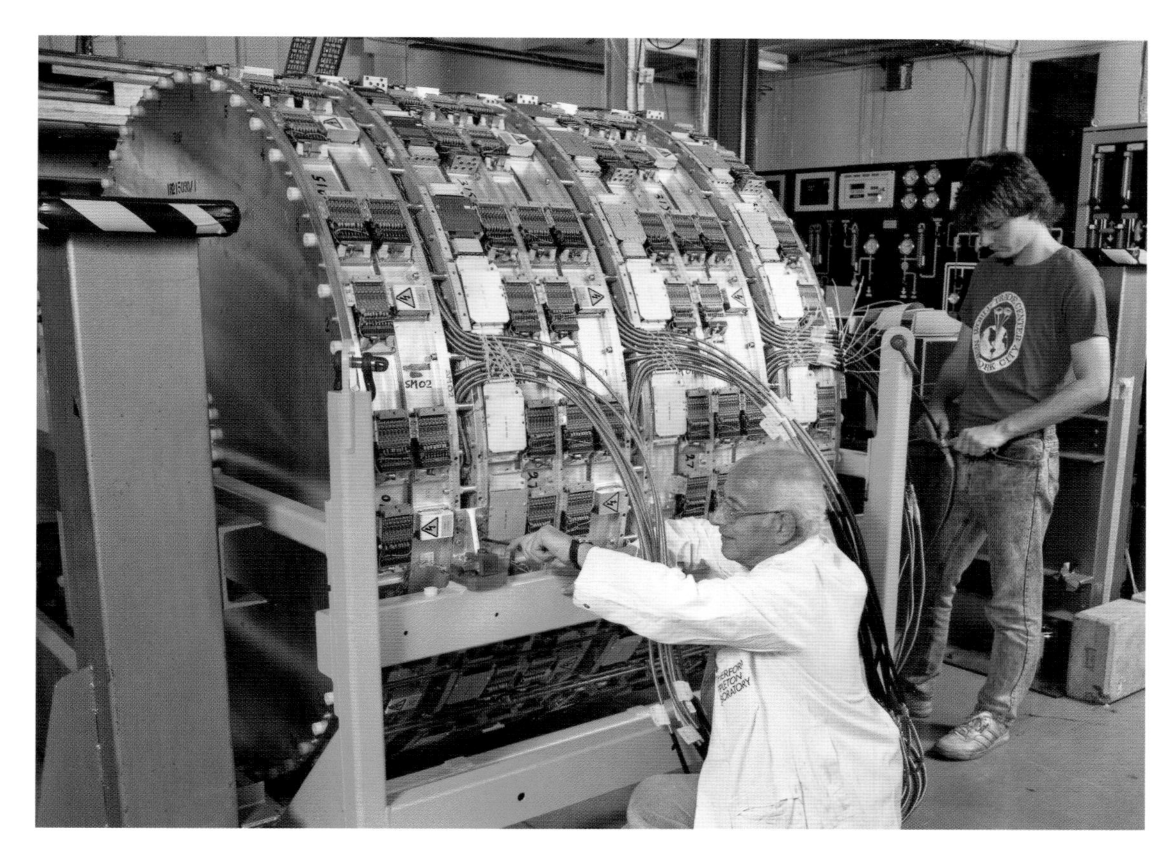

The Forward Track Detector (FTD) for the H1 experiment on the HERA (Hadron-Electron Ring Accelerator) machine at the DESY Laboratory under construction at RAL, 1990. The 6.3 km long storage ring facility HERA was used to accelerate protons and electrons. With the help of extremely sensitive building-sized detectors, like the FTD, it was possible to observe what happens when these particles collided with high energy. The H1 experiment was installed in 1992 to observe high energy particle collisions, allowing scientists to study the internal structure of the proton and the fundamental forces of nature. Fifteen years of scientific discovery came to an end this summer when electrons and protons in the HERA (Hadron-Electron Ring Accelerator) accelerator made their final lap of the storage ring.

A single supermodule of the FTD lying on its side in the cosmic test area, R12, 1990. The supermodule's high voltage system is just visible on the left and the gas system is on the right. The scintillators for the cosmic ray trigger are in the wooden supports.

H1 detector at DESY (Deutsches Elektronen-Synchrotron). The picture shows the H1 detector under construction. Credit: DESY

RAL open day, 1990. Thousands of visitors from around the UK were inspired by the science and facilities showcased during RAL's open day.

Professor Norman McCubbin showing school children Sophie and Bethan McDonald a particle physics event, 1992.

In April 1995, a new research council, the Council for the Central Laboratory of the Research Councils (CCLRC) was formed under Royal Charter to operate RAL and the Daresbury Laboratory (which had merged with RAL in 1994). RAL is now part of the Science and Technology Facilities Council, formed upon merger of the CCLRC with the Particle Physics and Astronomy Research Council in April 2007.

ELIZABETH THE SECOND
by the Grace of God of the United Kingdom of Great Britain and Northern Ireland and of Our other Realms and Territories Queen, Head of the Commonwealth, Defender of the Faith:

TO ALL TO WHOM THESE PRESENTS SHALL COME, GREETING!

WHEREAS it has been represented unto Us that it is expedient for the better execution of the purposes of the Science and Technology Act 1965 to make further provision for promoting and supporting scientific research:

AND WHEREAS Our Chancellor has appointed a person to be a member and proposes to appoint a Chairman and Chief Executive and other members of a Council for promoting and supporting scientific research to be known as the Council for the Central Laboratory of the Research Councils:

AND WHEREAS it has been represented unto Us that for the purpose of carrying out the objects of the said Council and with a view to facilitating the holding of and dealing with property and to encouraging the making of gifts and bequests in aid of the said objects it is expedient that the said Council should be incorporated:

NOW THEREFORE Know Ye that We, by virtue of Our Prerogative Royal and of all other powers enabling Us so to do, have of Our especial grace, certain knowledge and mere motion granted and declared and do by these Presents for Us, Our Heirs and Successors, grant and declare as follows:

1 The person appointed by Our Chancellor to be a member, and all such other persons as may hereafter become the Chairman and Chief Executive and other members of the body corporate hereby constituted, shall for ever hereafter (so long as they continue to be members of the Council) be one Body Corporate under the name of "The Council for the Central Laboratory of the Research Councils" (hereinafter referred to as "the Council"), and by the same name shall have perpetual succession and a Common Seal, with power to break, alter and make anew the said Seal from time to time at their will and pleasure and by the same name shall and may sue and be sued in all courts and in all manner of actions and suits, and shall have power to enter into contracts, to acquire, hold and dispose of property of any kind, to accept trusts and generally to do all matters and things incidental or appertaining to a Body Corporate.

2 (1) The objects for which the Council is established and incorporated are:

(a) to promote high quality scientific and engineering research by providing facilities and technical expertise in support of basic strategic and applied research programmes funded by persons established in this Our United Kingdom and elsewhere;

(b) to support the advancement of knowledge and technology, meeting the needs of research councils, other customers and their user communities, thereby contributing to the economic competitiveness of Our United Kingdom and the quality of life;

Sir Tim Berners Lee (right) receiving a medal at the 1997 World Wide Web Consortium (W3C) Meeting. In 1989, to make life easier for scientists, Tim invented the World Wide Web, an internet-based hypermedia initiative for global information sharing while at CERN. He wrote the first web client and server in 1990. Tim's specifications of URLs, HTTP and HTML were refined as Web technology spread.

CMS before the detector was closed, 2006. The ATLAS and CMS detectors, being built for the Large Hadron Collider (LHC) at the CERN Laboratory in Switzerland, will explore the fundamental nature and the basic forces that shape our universe. The detectors, an order of magnitude larger and more complex than any previous experiments, will have to be capable of precisely measuring the ~40 charged particles emerging from each of the thousand million collisions occurring every second.
Credit: Patrice Loiez, CERN

Julian Williams checking components of the forward/backward sections of the electromagnetic calorimeters under construction for the CMS detector, 2004.

Scientists dwarfed by the ALEPH detector, CERN, 2004.

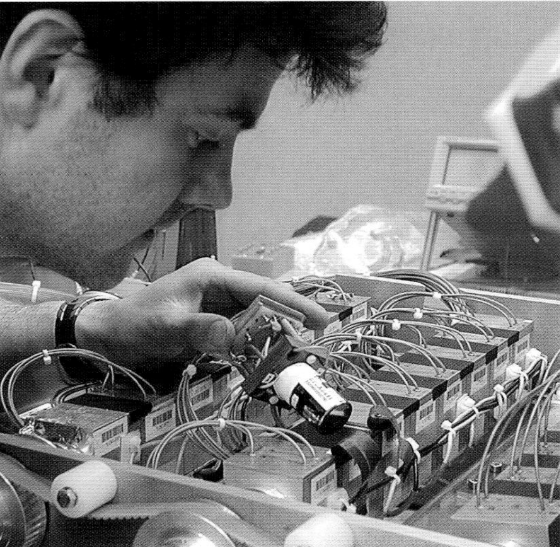

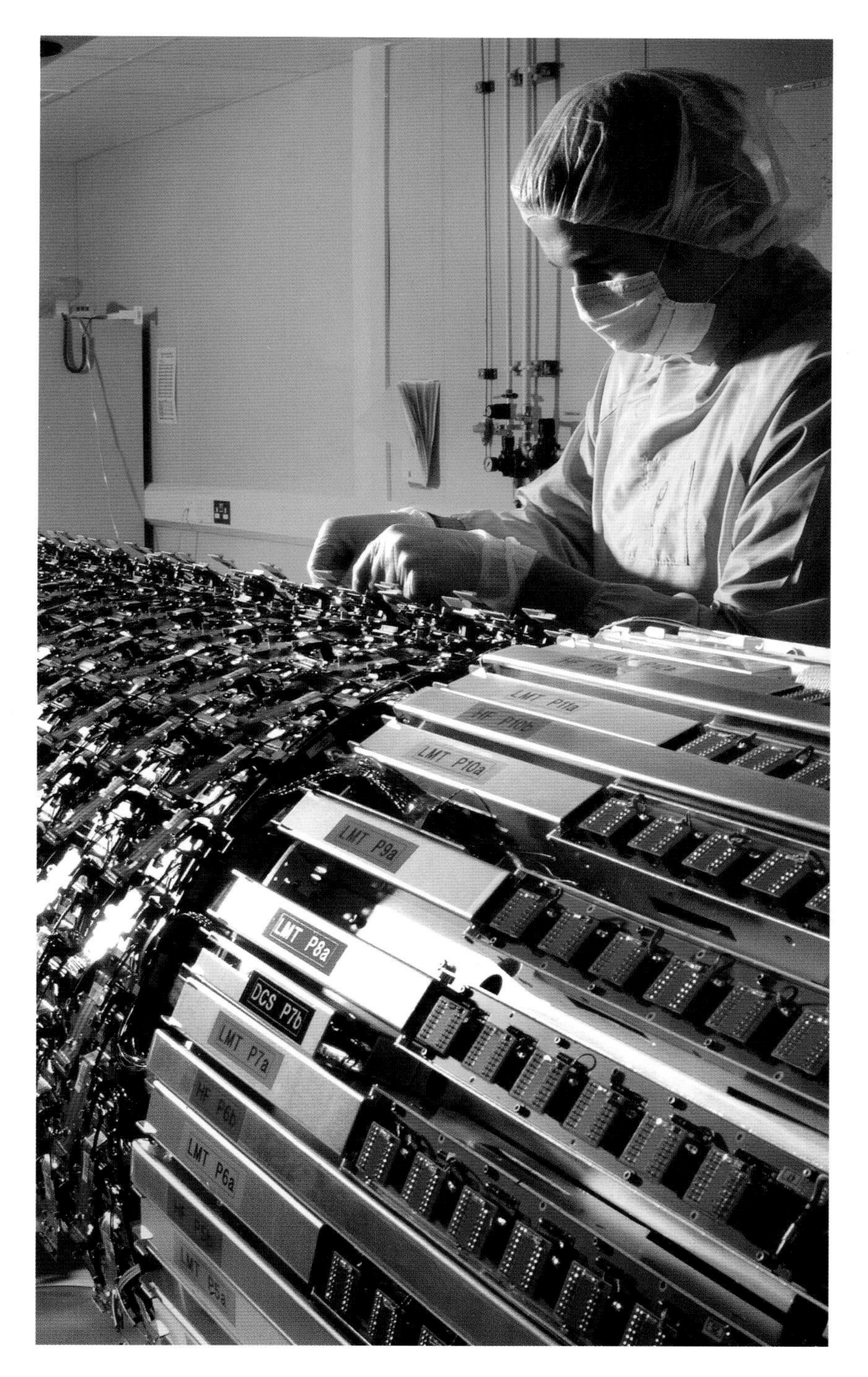

Impressive views of ATLAS cavern side A, 2007. The detector module built at RAL is one of the principal elements of the ATLAS inner detector.

ATLAS detector module under construction, 2004. Lewis Batchelor making some final adjustments in the R12 clean rooms.

A detailed view of the electronics on the ATLAS Semi-Conductor Tracker (SCT) end cap, 2004. Constructed at RAL, the SCT for the ATLAS detector is the largest particle physics project ever undertaken in the UK.

An A-level student from Downe School explores the structure of materials whilst touring ISIS during RAL's 2004 Particle Physics Masterclass. The masterclass is a popular series of one day events for sixth form students and their teachers, run by practising particle physics researchers at the laboratory. RAL's Particle Physics Masterclasses provide excellent support for the material on Particle Physics which is in many 16+ curricula.

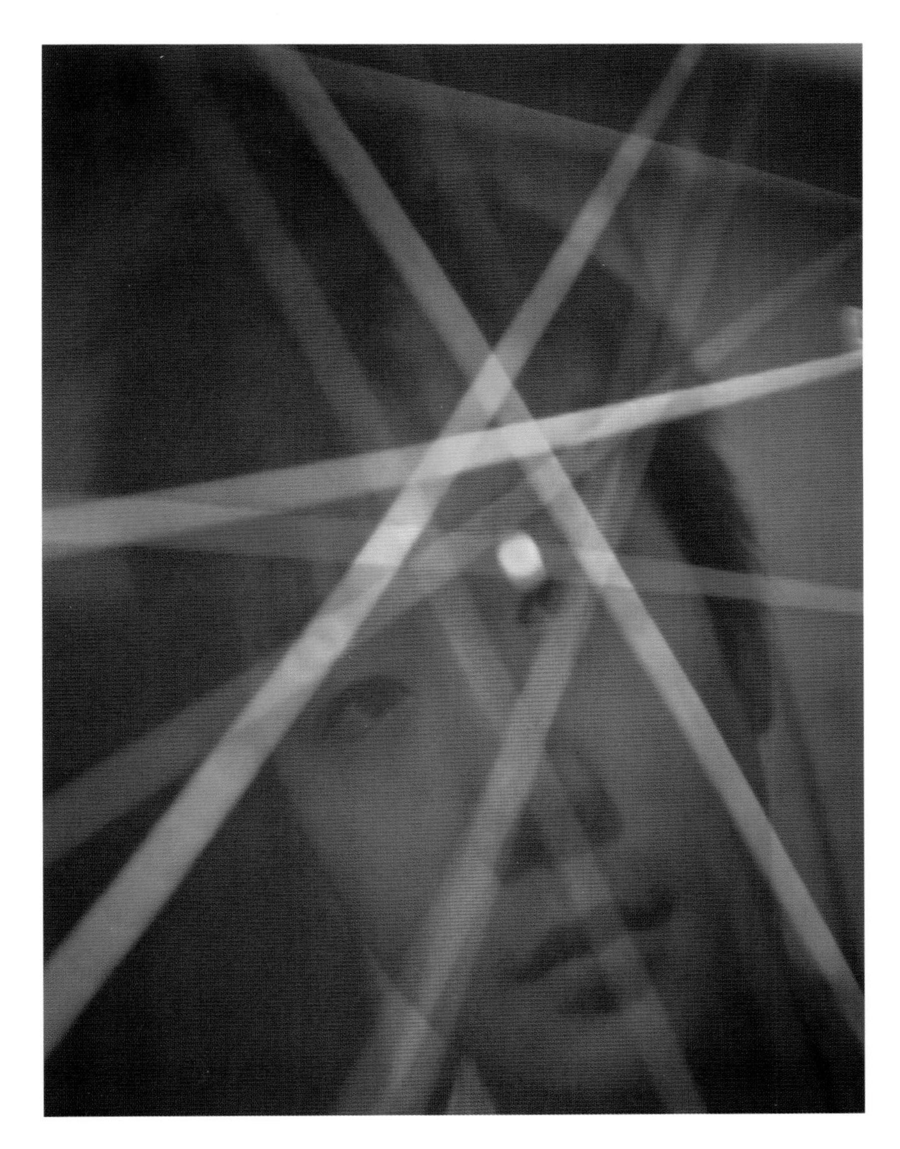

The Particle Physics Masterclasses has proved so popular that RAL now runs three masterclasses per year. Thousands of students have been educated, stimulated and inspired through the courses since they began in the late 1990s.

Dr Emmanuel Olaiya with students in RAL's training room during the 2006 Particle Physic Masterclass.

Technology

An 8 inch wafer of CMOS sensors designed at RAL for an electron imaging application. The sensors each contain an array of 25 (5 by 5) layouts that can be seen in each tile.

World-leading science facilities require world-leading technology. As a consequence the Rutherford Appleton Laboratory has invested heavily over the past 50 years in developing support technologies which have benefited a number of scientific fields, and society as a whole. RAL has also assisted the movement of technologies across scientific disciplines, and more recently, accelerated the transfer of these technologies to the commercial sector.

Many of these accomplishments have stemmed from the extreme demands made by the design of particle physics experiments. The laboratory was a pioneer in the development of superconducting magnet technology in the early 1970s. This technology was then used in the world's largest superconducting solenoid for particle physics (DELPHI) and transferred to the emerging science of synchrotron radiation through the production of the first high-field wiggler for the Synchrotron Radiation Source at Daresbury Laboratory.

RAL's detector technology is based on in-house capability to design entire 'detector chains' from the sensing elements, through novel microelectronics read-out chips to large scale data acquisition systems and display software. This expertise is now deployed over a range of applications from space science and neutron physics to biology and brain science.

The programme in space science has also led to many significant and useful technologies. As an example, the development of waveguide technology for millimetre-wave radio astronomy began at the Appleton Laboratory in the 1970s, and was applied to the UK Infrared and James Clerk Maxwell Telescopes on Hawaii. Spin-out technology based on programmes for the detection of terahertz radiation has led to the development of real-time cameras which are able to image potential security threats such as guns or explosives carried by airline passengers.

At the Rutherford Appleton Laboratory the technology base was developed to meet the needs of particle physics. It has grown dramatically over the past 50 years, underpinning all the science programmes and delivering economic benefits through spin-out and knowledge exchange.

An engineer checking the alignment of an instrument for Nimrod, 1958. Engineers, instrumentation scientists and technologists have underpinned activities at RAL since the Chilton site's development first began in the 1950's.

Professor Ron Lawes, former Head of the Central Microstructure Facility, using the HPD scanning machine laser, 1970.

With a diameter of 15 m the James Clerk Maxwell Telescope (JCMT) is the largest astronomical telescope in the world designed specifically to operate in the submillimeter wavelength region of the spectrum. Designed and built by a Dutch-UK collaboration, the JCMT won the McRobert Engineering Award in 1987. The JCMT is used to study our Solar System, interstellar dust and gas, and distant galaxies. It is situated close to the summit of Mauna Kea, Hawaii, at an altitude of 4092 m.

The surface of JCMT's main dish consists of 276 panels, each of which can be moved by means of three motorised adjusters. The panels were designed and fabricated at RAL and consist of honeycomb structures with thin aluminum surface panels glued on top.

The world's largest superconducting solenoid under construction at RAL, 1987. A key component of the Delphi detector, the superconductor was so large that it had to be transported to Switzerland by road and sea.

Delphi was a Particle Physics experiment at the CERN laboratory in Geneva, Switzerland. Delphi studied the products of electron-positron collisions at the Large Electron-Positron (LEP) circular accelerator, working at the highest energies in the world. Operation of Delphi started in August 1989 when the LEP was switched on and stopped in November 2000 with the closure of the LEP facility.

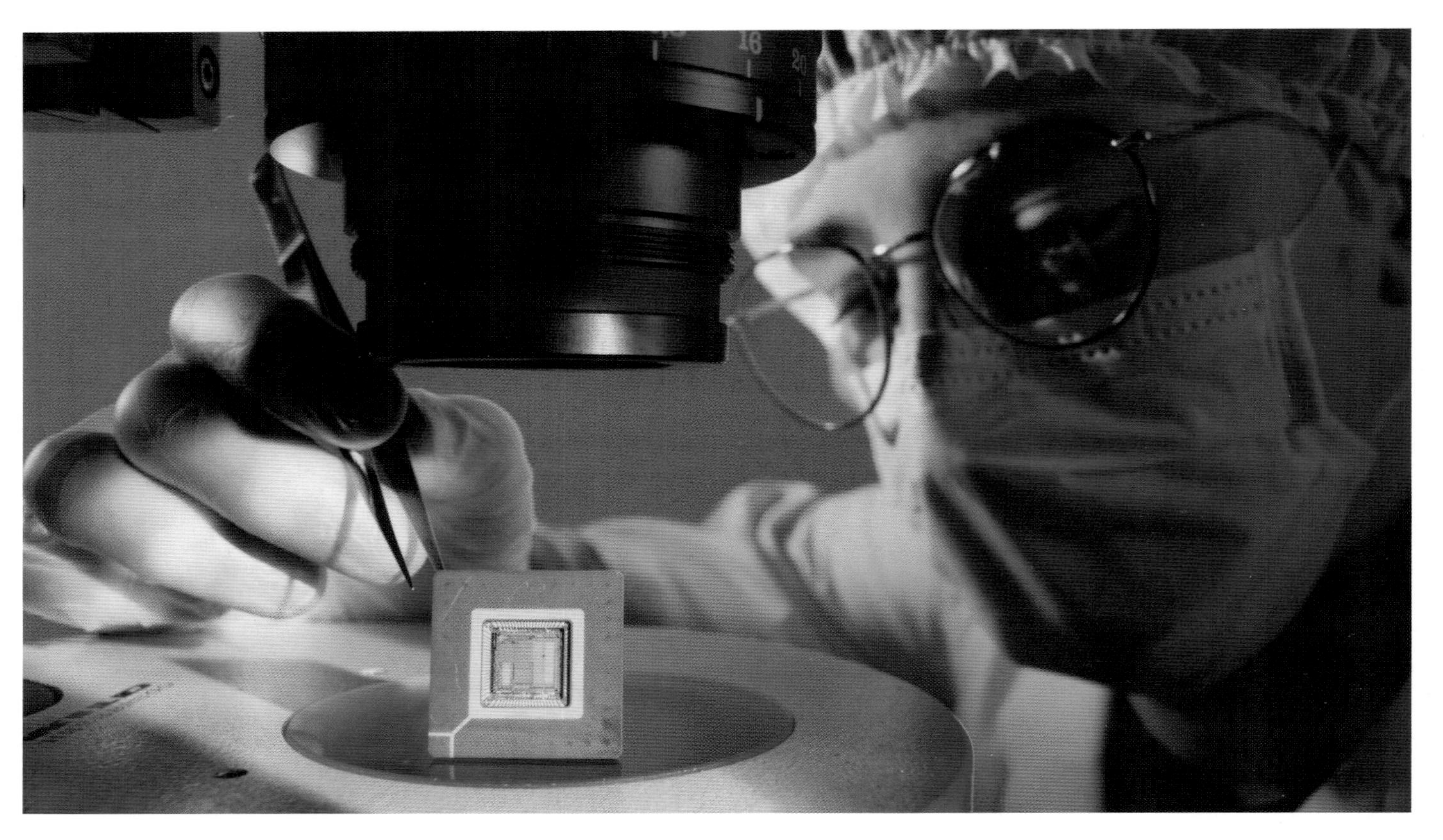

An apprentice inspecting a microchip made at RAL, 1993.

Electron Tubes, a spin-out from RAL, launched a new product at the Photonics West show in 2001 based on fast data acquisition chips for X-ray readout. The chips, originally designed at RAL for CERN detectors, have potential applications in steel strip measurement, bone density assessment, food X-ray and security screening.

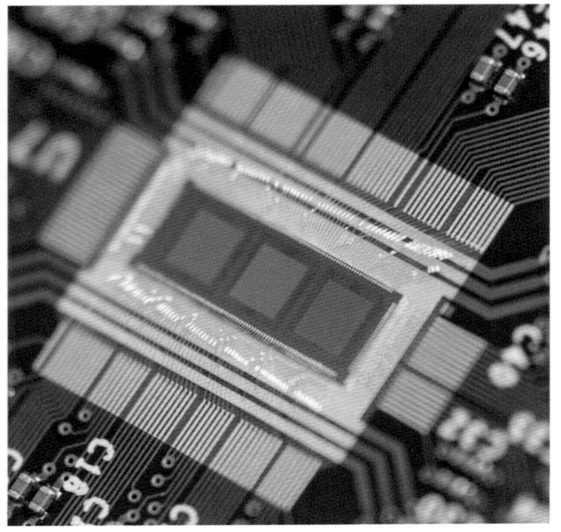

The Microelectronics Design Group is responsible for the design and development of state-of-the-art analogue electronics for readout of the most sensitive detector systems used by the international research community.

A prototype pixel sensor which has potential benefits for high frame-rate imaging and applications using lasers or particle physics, 2006. RAL's CMOS Sensors Group develops image sensors for a wide range of customers using industry-standard CAD tools.

RAL Test Engineer Craig MacWaters looking underneath the chips on a front-end driver (FED) board, 2006. The Compact Muon Solenoid (CMS), under construction for the LHC at CERN, employs the world's largest silicon-strip detector. RAL is developing the readout system which has to acquire data from 10 million silicon strips at a rate of up to 100,000 times a second.

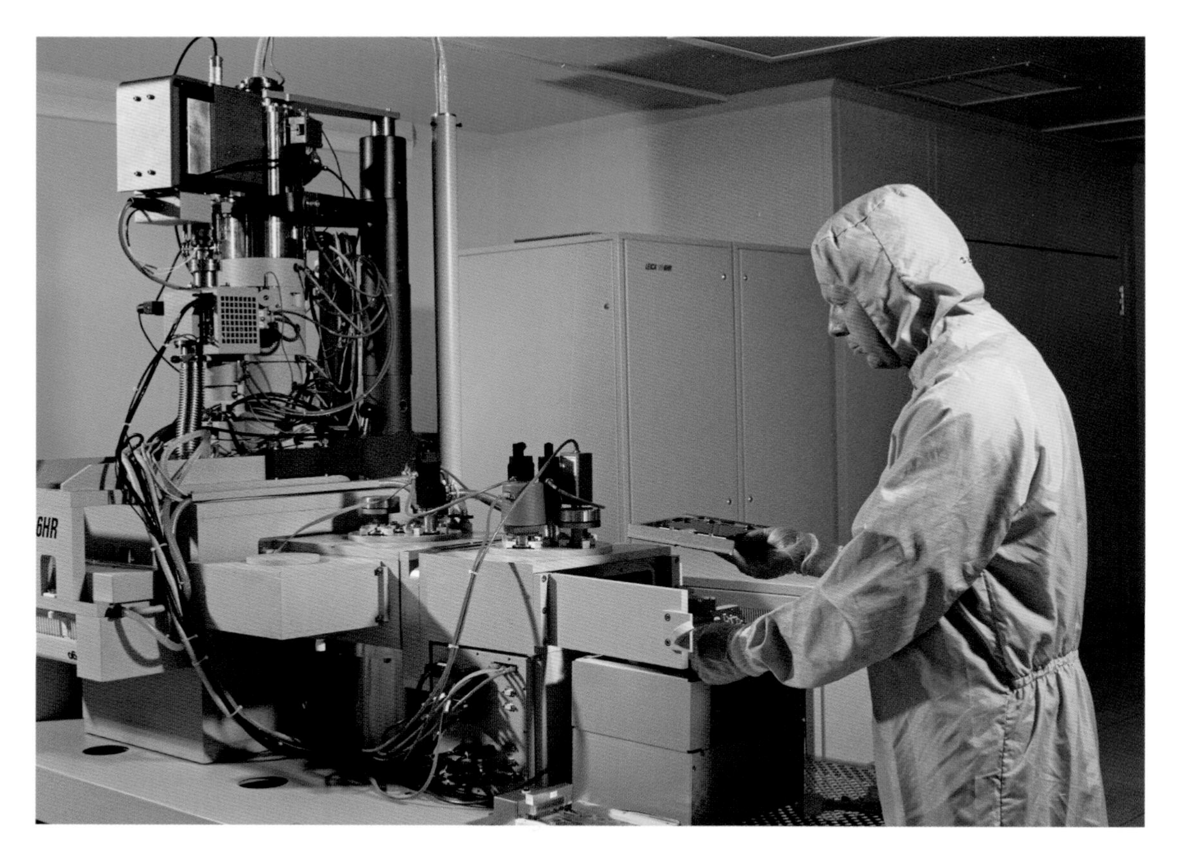

Ian Loader at work on the Electron Beam Lithography machine in the CMF, 1996. The Leica VB6-HR electron beam lithography tool installed in the clean rooms of the Central Microstructure Facility at RAL. Its highly focussed electron beam can be steered under computer control to define the micro- and nano-scale features found on photomasks, silicon wafers and other devices. This machine was used to write the smallest advert in the world (see below).

In 1999/2000 the CMF took up the challenge to produce the world's smallest advert to advertise the launch of the new Guinness World Records' website. The result was an advert so small that it could fit on a bee's knee.

A nickel gear wheel produced by the Central Microstructure Facility (CMF) in 2001 (pictured on the end of a finger). The wheel was produced using the precision micro-engineering technique X-ray LIGA (Lithographie, Galvoformung und Abformung), i.e. lithography, electro-deposition and moulding. In addition to chrome-on-glass photomask manufacture, the CMF produces gold-on-polyimide and gold-on-beryllium masks for the technique.
Credit: Graham Arthur

Silicon tip array with tip apex radius of sub-10 nm created in the CMF.

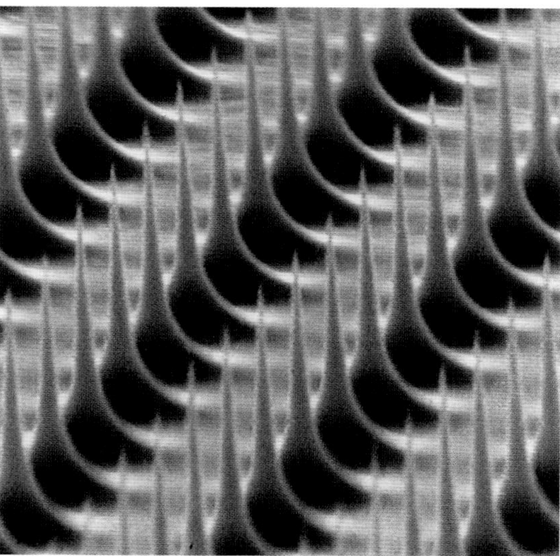

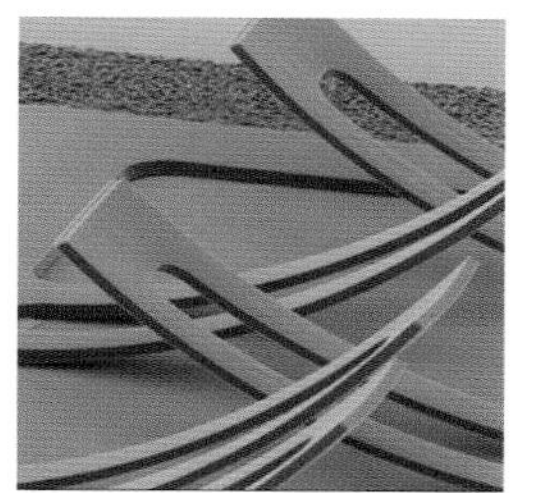

Microvisk, a company spun-out from the Central Microstructure Facility by RAL's knowledge transfer company, CLIK, in 2001, is exploiting a novel microfabricated approach for viscosity and viscoelastic measurements which has a multitude of potential applications. Part funded by the Rainbow Seed Fund, Microvisk is currently designing desk-top equipment that can quickly and accurately measure blood viscosity.

Application Specific Integrated Circuit on a silicon wafer, 2000. RAL has unique skills in the design and modelling of both analogue and digital ASICs in three dimensions, allowing for subtle feature size, timing and power consumption effects that come into play in the ultra small chip design regime. ASICs allow much of the electronics in space cameras to be miniaturised and placed within a small set of integrated circuits. RAL's ASICs are unique in that they are designed specifically to provide high imaging performance whilst withstanding the harsh environments of space.

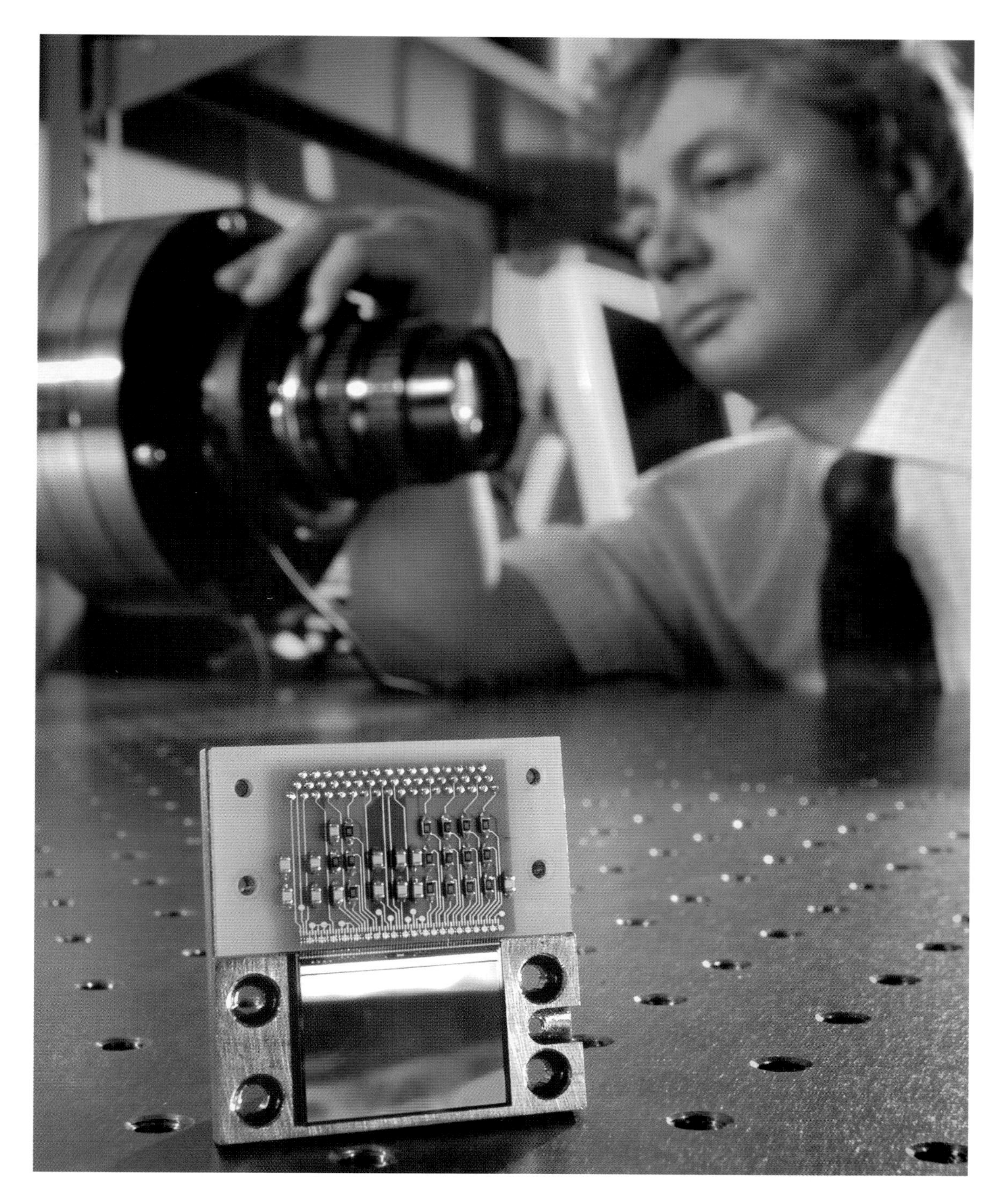

With the 12 million pixel sensor in the foreground, Nick Waltham adjusts a test camera for ESA's Solar Orbiter EUV Spectrograph, 2004. RAL developed CMOS (complementary metal oxide semiconductor) Active Pixel Sensor (APS) technology for ESA's Solar Orbiter which is due for launch at the end of the decade. Solar Orbiter will experience extreme particle environments precluding the use of traditional CCD detectors. Sensitive to a wide range of radiation, APS sensors have many potential uses, for example, as a star tracker in space science applications or to detect charged particles in high-energy physics experiments

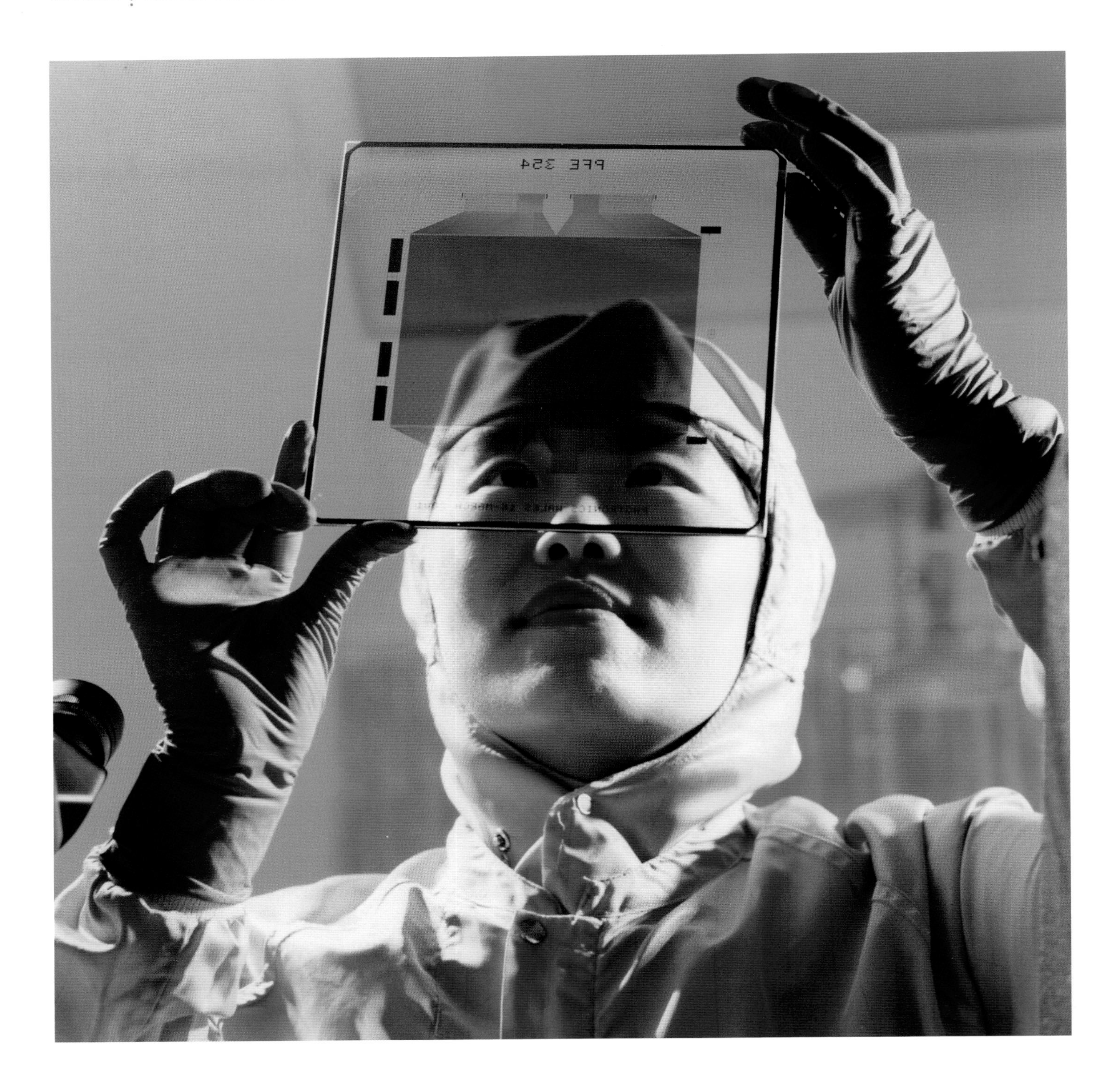

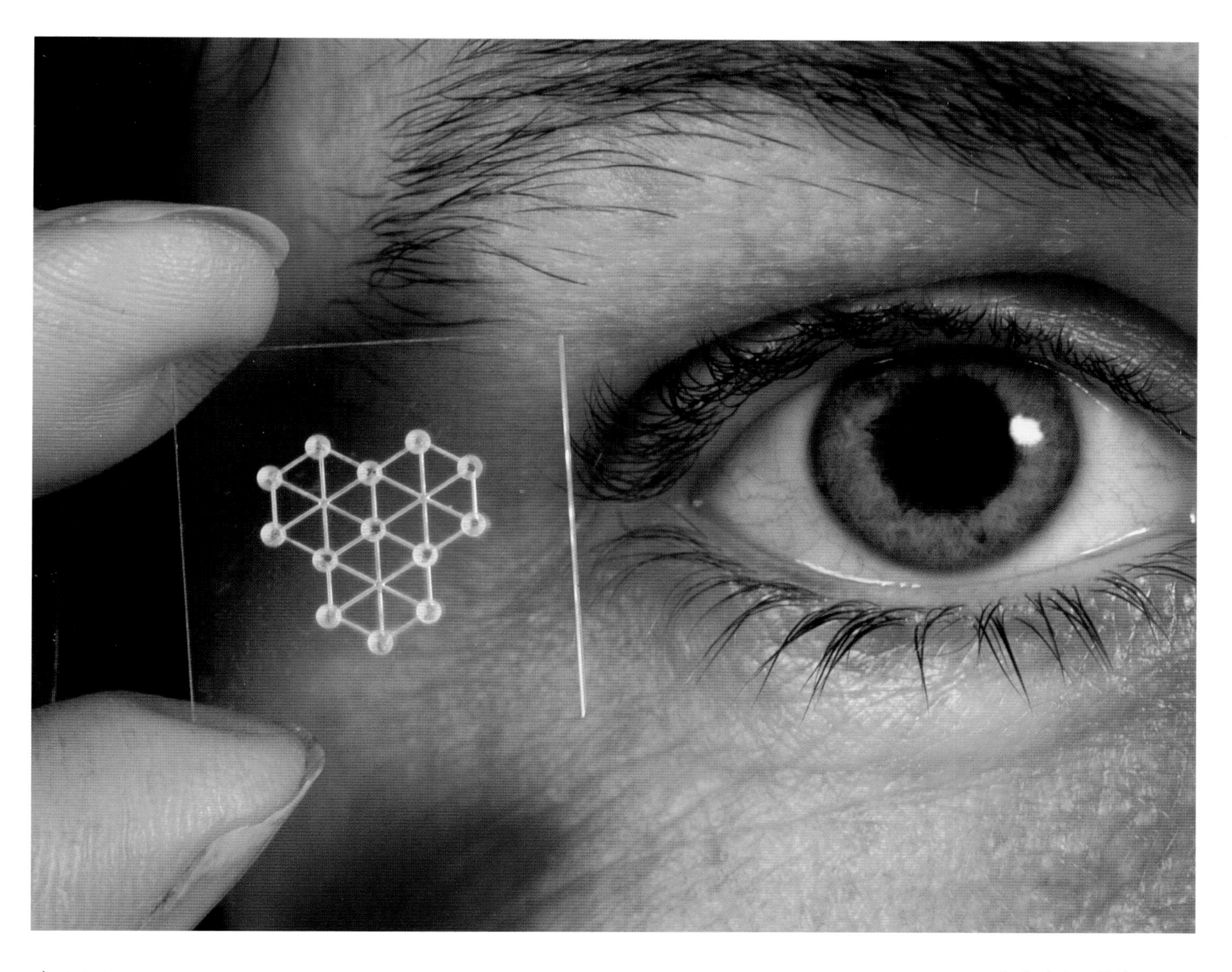

The high technology spin-out company Printable Field Emitters Limited (PFE Ltd) developed a radically less expensive way to manufacture Wall TVs using techniques developed in the CMF clean rooms, 2001.

A minute scaffolding manufactured by the CMF as part of a project, with teams from the universities of Nottingham and Leeds, to create an artificial liver, 2004. The structure simulates the complex tubes that occur naturally and which carry the blood around the liver.

Precision measurement technology is at the heart of RAL's Metrology Facility. Each component built at RAL undergoes extensive measurement and quality control processes throughout its development cycle. Here the holes drilled in an ATLAS detector module are measured using state-of-the-art technology, 2006.

Lewis Bachelor working on one of the ATLAS detector modules for the Large Hadron Collider, 2004. Each detector had to be built to very exacting standards, for example, each silicon sensor had to be located to an accuracy of better than 5 microns, and then each of the 1536 separate readout channels was individually wire bonded to the read out electronics. Typically, there were 40 modules in various stages of construction at any one time with each module taking two weeks to complete.

Arnold Harpin of Oxsensis, 2004. Oxsensis is applying RAL technology for sensors in extreme environments such as jet engines and giant transformers on the National Grid. Spun-out from RAL, Oxsensis is developing sensors that can face temperatures up to 1000 degrees C, electrical discharges and extreme ultraviolet light. Oxsensis attracted new funding of £4.36 million in 2007.

C.M.F.

The 10 ton rig being used to mechanically test components at very low temperatures using liquid Nitrogen as the coolant in an unsilvered Vacuum Flask at RAL, 2007. The suitability of materials used to manufacture research equipment and facilities and their ability to cope with extreme environments is tested rigorously using on-site expertise.

Dave Wilsher in the Metrology Facility, 2006.

The dimensions of a metal cube being measured with a ruby probe on a CMM machine, 2007. Precise measurement techniques and equipment are constantly updated, keeping RAL at the cutting edge of technology.

The Energy Research Unit's Windharvester Wind Turbine on RAL's test site facility, 2007. The Energy Research Unit at RAL has been carrying out front-line energy research for over 25 years and has many collaborative links with academia and industry both in the UK and abroad. Wind and solar energy have been studied at RAL for many years. Current research includes fuel cells, fusion energy, photovoltaics, renewable energy and hydrogen storage.

Reflection of clouds in a solar panel on the test site facility, 2000.

Engineers of the future learning about wind energy during a National Science Week workshop, 2001.

An apprentice examining a devise via non-contact measuring, 2002. RAL has been offering a number of electrical, electronic and mechanical apprenticeships since the 1980s. Following the completion of their training, many apprentices have chosen to further their career at RAL. The apprenticeship scheme is now regarded as one of the best engineering apprenticeships in the country.

An ingenious tiny 'puzzle' of cubes, machined from a single piece of Perspex. This fascinating object was made by one of RAL's engineering apprentices, 2005.

Students at the Engineering Education Scheme residential workshop, 2006.

Computing

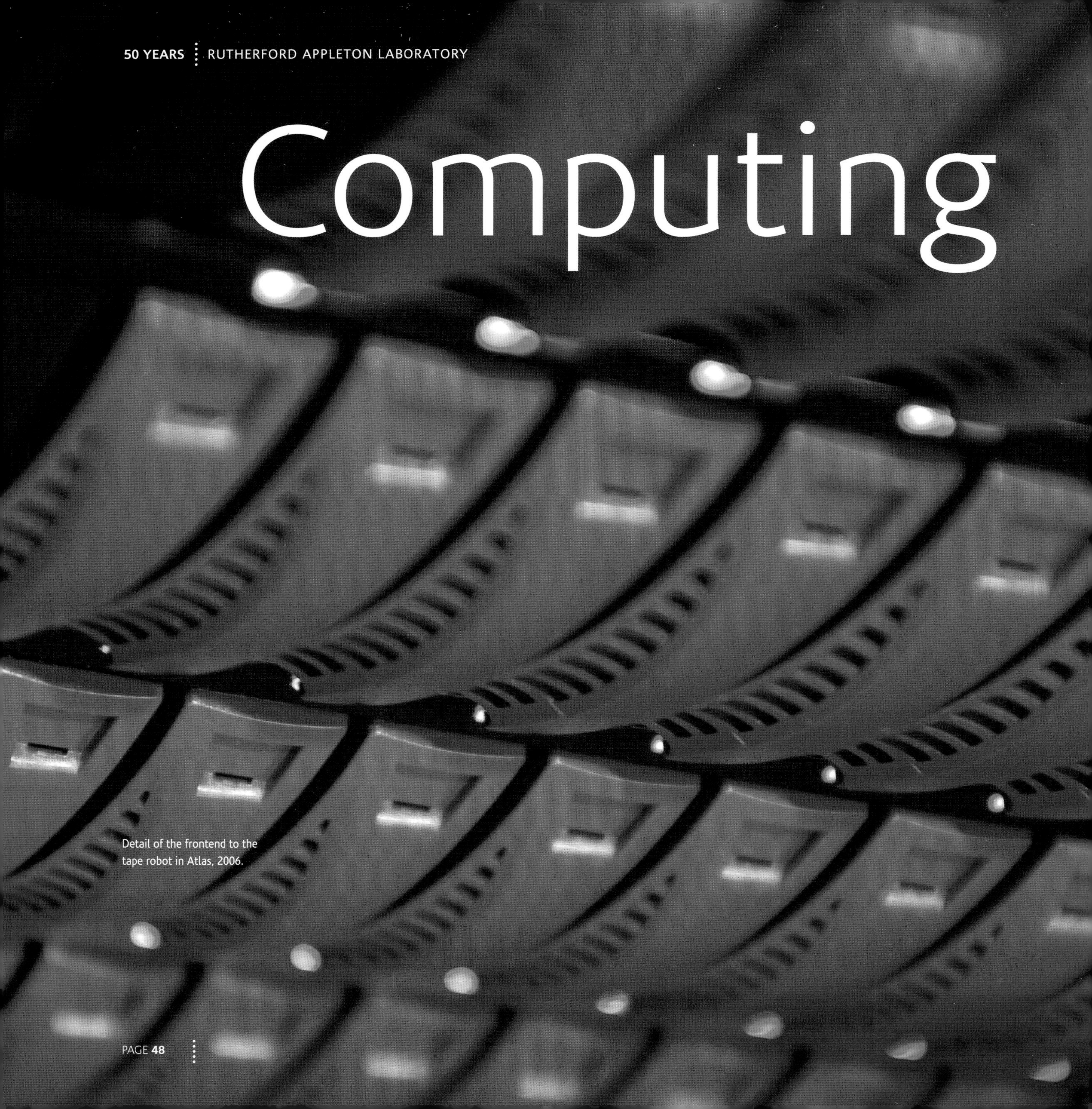

Detail of the frontend to the tape robot in Atlas, 2006.

The Atlas Computing Laboratory was established at Chilton in 1964 to support UK university research in the sciences, social sciences and humanities. Staff were drawn from the Government Code and Cypher School and Harwell. In the same year, Atlas installed the world's most powerful computer, the Ferranti Atlas. World leading applications in numerical analysis, text management, database technology and graphics were developed. To maintain its cutting-edge service, the more powerful IBM 360/195 was installed by Atlas in 1971. Two years later, the IBM 360/195 became the first mainframe outside the USA to connect to the ARPAnet.

The laboratory was renamed the Atlas Centre in 1975 and merged with the neighbouring Rutherford Laboratory. The same year brought the establishment of SRCnet, linking the IBM at Rutherford with one at Daresbury Laboratory and the ICL 1906A at Atlas. This network was extended in 1984 to become the first UK national computing network, JANET. JANET was privatised as UKERNA in 1994.

1987 saw the arrival of the first supercomputer at RAL, the Cray X-MP, providing 1 Gflop, followed by the even more powerful Cray Y-MP (2.5 Gflops) just five years later.

One of the first web servers in the world was installed at RAL in 1992 and the W3C (World Wide Web Consortium) Office for UK and Ireland continues to be hosted at RAL. Networking has continued to advance, offering more computing power to the scientific community. In 2004, the National Grid Service became the first operational national grid service in the world while only last year, the RAL Tier-1 computing service broke a world record through sustaining a 200 Mbyte/s data transfer with CERN.

RAL's computing facilities and staff support the scientific community, providing massive data storage and processing facilities, and enabling the development of effective disaster management systems, multimedia systems, data and knowledge management systems, formal proof software systems, visualisation systems, communications systems and the preservation of digital data. One of the most recent successes has included biological visualisation, with applications from testing new drugs to planning clinical interventions.

Geoff Lambert at the console of the IBM 360/75 computer, 1967.

Animators John Hallas (Animal Farm), and Stan Hayward (Henry's Cat) worked with the Laboratory on a number of projects. Through them, Atlas was involved in making a Tomorrows World programme about the potential of computer animation. Professor Bob Hopgood animated a walking man, shown by the BBC, using the GROATS animation facilities. 1971.

Friden

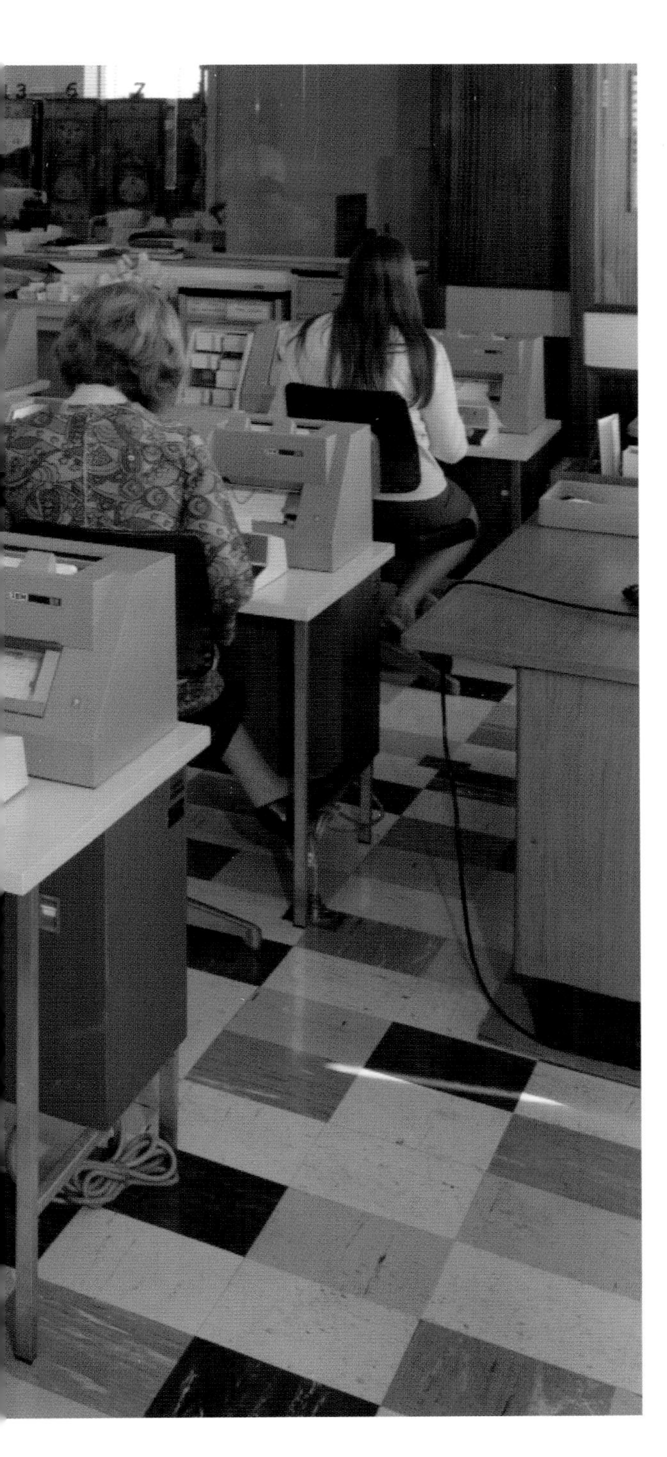

VDU operator Ruth Jeans, 1976. Ruth is now a member of the Council Secretariat.

Atlas computer centre, R27, 1965.

Directors Dr Jack Howlett and Dr Brian Davies in the Atlas display area with the control panels from the IBM 360 and Atlas computers, 1984. The nameplate of the Atlas locomotive is on the wall.

Cray X-MP vector supercomputer arrives in Atlas, 1984. The Cray X-MP, a parallel vector processor machine was the world's fastest computer 1983–1985.

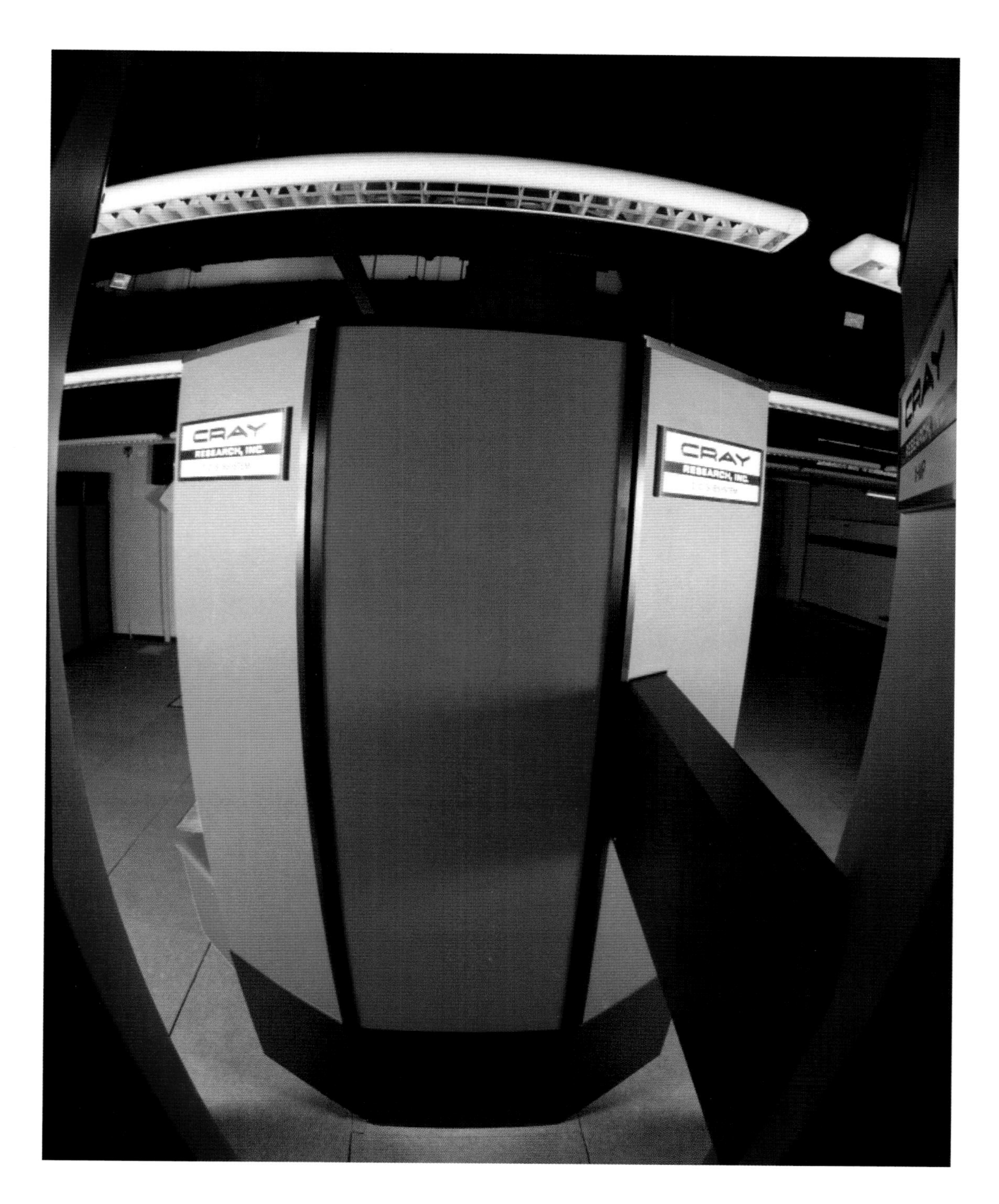

ATLAS CENTRE
JOINT RESEARCH COUNCILS
SUPERCOMPUTER UNIT
CENTRAL COMPUTING DIVISION
RUTHERFORD APPLETON LABORATORY

Dr Brian Davies, head of the Central Computing Department, testing the new Cray Y-MP computer which was operated on behalf of the UK Research Councils, August 1992 .

Atlas in springtime, 1991.

Dr Paul Williams, Laboratory Director and Professor Sir Mark Richmond, Chairman of the Science and Engineering Research Council (SERC), look on as William Waldergrave, Chancellor of the Duchy of Lancaster runs a program to mark the official inauguration of Cray Y-MP, 22 April 1993.

Martin Prime and Rachel Miles in The Engineering Applications Support Environment (EASE), 1993.

Dr Lakshmi Sastry with Nuffield Bursary Student Emma Quelch, who has now received a PhD herself, in the Virtual Reality Laboratory, 2001.

David Boyd explores the proposed ATLAS magnet for CERN in virtual reality in the R27 Virtual Reality lab, 1996.

The Access Grids at RAL, 2003. The Access Grid project has established Access Grid nodes (rooms) at Rutherford Appleton Laboratory and, sister laboratory, Daresbury Laboratory modelled on nodes in the US. They provide a 20' x 5' wall filled with images from three video projectors. Several video feeds (normally 4) are sent from each room in the meeting, to one or more attending sites, allowing it to be used for meetings between several groups. Access Grid also provides 'tools' to share presentations and web browsers, providing a more effective way of working together with little or no travel overhead.

Inside the tape robot of the Atlas Petabyte Store, 2006.

Hiten Patel in the Atlas Petabyte Store, 2006. The digital data store provides 5 Petabytes of on-line storage. The APS service provides very large amounts of permanently accessible digital storage. Used by many users throughout the UK research community, the service has been developed significantly in preparation for the start up of the Large Hadron Collider experiment at CERN in 2008.

The 2007 VizNet Showcase first prize was awarded to the e-Science centre's Dr Lakshmi Sastry for her distributed, high resolution visualisation of a rabbit heart using commodity cluster and open source software stack. The visualisation was developed as part of the Integrative Biology Project. The project is bringing together an international consortium of leading biomedical and computing researchers to address two of the most important problems in clinical medicine today: understanding what causes heart failure and how cancer tumours develop and grow.

ISIS

Sub-atomic particles called muons can be implanted into materials where they exist fleetingly as tiny 'spies' that inform us about the atomic-level details of the material. Here, a muonium atom (a positive muon that has picked up an electron to form an atom resembling hydrogen) is trapped inside a C60 cage (a 'buckyball').

Credit: Created by Robert Dalgliesh with input from Kenneth Shankland and Adrian Hillier.

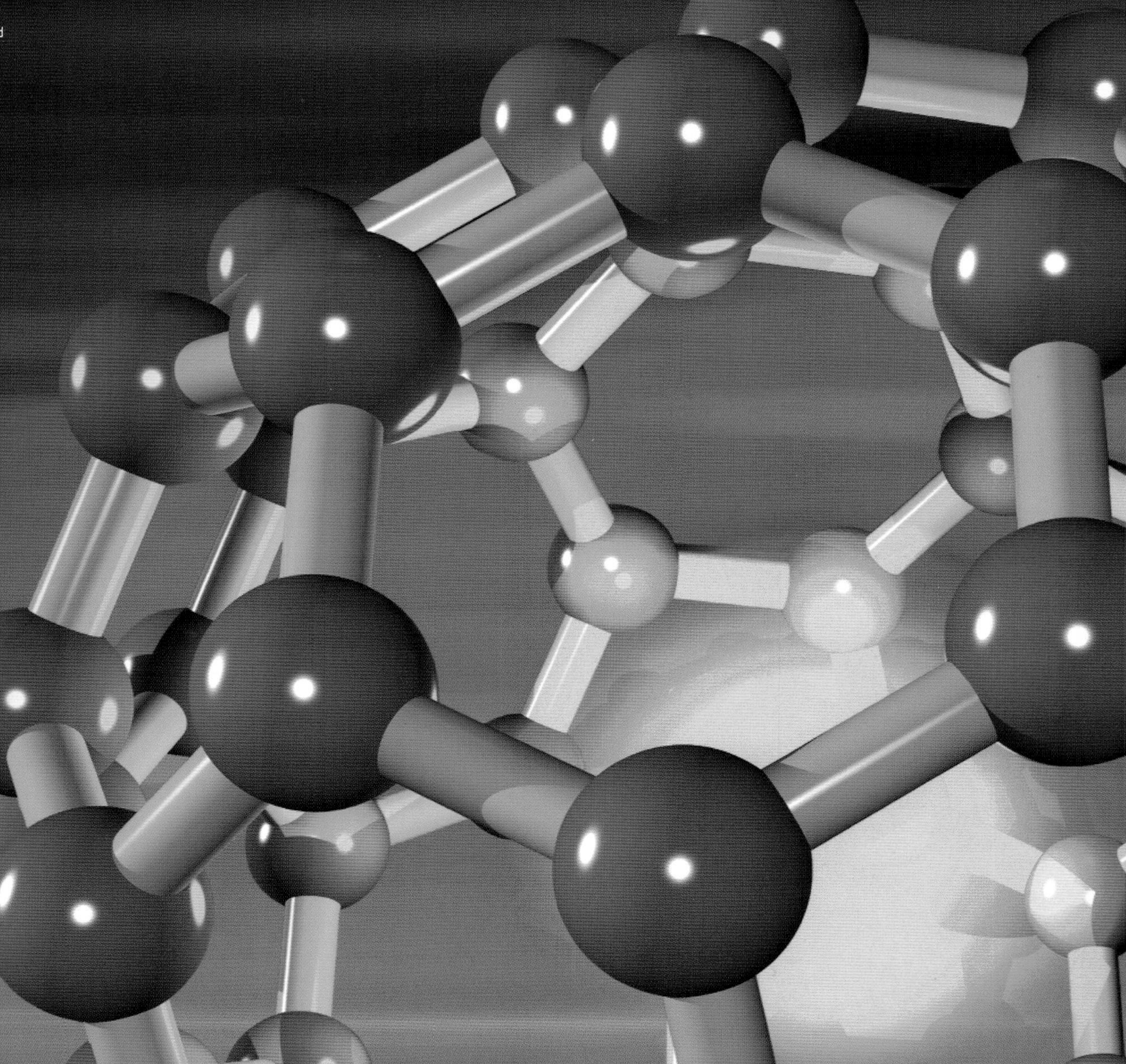

A confluence of capabilities and opportunities in the mid 1970s – the right people in the right place at the right time – allowed Rutherford Appleton Laboratory to develop ISIS, currently the world's leading pulsed neutron source.

ISIS emerged as a phoenix from the UK domestic particle physics programme as Nimrod came to the end of its productive life. Utilising advanced skills in accelerator design and project engineering, and drawing on inherent competences in instrumentation, detector technologies and data acquisitions systems, ISIS developed into a world leading facility for condensed matter science.

Based on an innovative, high intensity, rapid cycling synchrotron, ISIS recycled many components – the Nimrod Cockcroft-Walton pre-injector, two tanks from the Proton Linear Accelerator, the Nina choke donated by Daresbury Laboratory, an extensive experimental hall, and much valuable shielding – to create a world ranking facility for the UK at an affordable cost.

Many played their part. Leo Hobbis, George Stirling and Harold Wroe from the Laboratory, and Bill Mitchell and Alan Leadbetter from the research community all played leading roles, along with many others. But the inspiration was that of Geoff Manning.

The rest is history. A young enthusiastic team of neutron scientists, building on the technologies developed in the Laboratory and elsewhere, created a science driven culture which, working intimately with the community, delivered a world leading facility that defined the future direction for neutron scattering world wide. In the past 20 years, it has given its academic and industrial communities – both from the UK and overseas – unique capabilities to address key problems in physics, chemistry, biology, engineering and materials science.

A second target station, which will increase the experimental reach and double the capacity, is nearing completion. The extended capability will allow ISIS to expand into areas of advanced materials, soft condensed matter, biomolecular systems and nanoscience, with the possibility of future upgrades to multimegawatt capability. ISIS is a real jewel in the crown of UK science.

ISIS under construction, 1982. Construction of ISIS began in 1978 when Nimrod was switched off and dismantled.

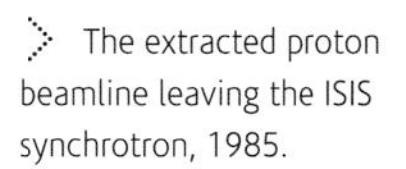

The extracted proton beamline leaving the ISIS synchrotron, 1985.

Beam on! The first neutrons were produced in late 1984 and ISIS was officially inaugurated on 16 October 1985.

Aerial view of ISIS, 1982.

Flags of the international partners fly outside the ISIS facility, 1996.

Professor Alan Leadbetter instructs Dr Andrew Taylor on how to analyse the first spectrum taken on ISIS during commissioning run! 1984.

Dame Julia Higgins, former Chairman of EPSRC, was one of the early advocates of ISIS (photo courtesy of James Hunkin).

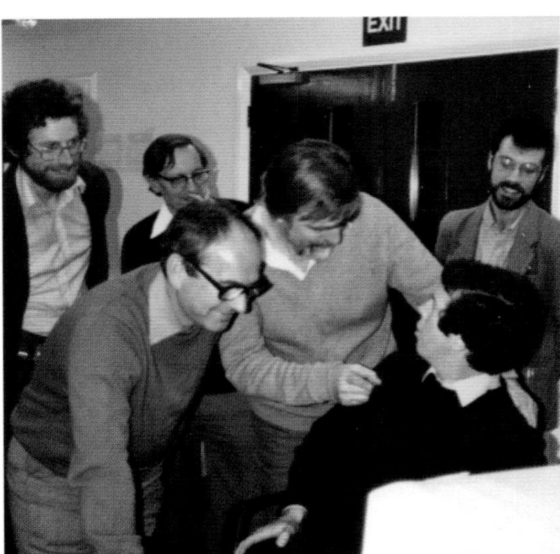

KARMEN neutrino facility at ISIS, 1988. Karlsruhe-Rutherford Medium Energy Neutrino (KARMEN) was an experimental neutrino physics programme investigating charged current and neutral current neutrino nucleus interactions.

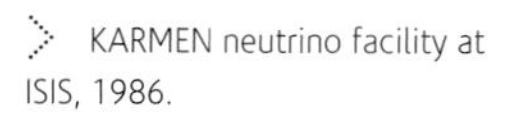

KARMEN neutrino facility at ISIS, 1986.

ISIS Experimental Hall with the RIKEN Muon facility in the foreground, 1994.

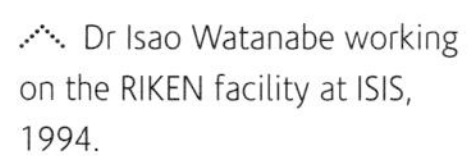

Dr Isao Watanabe working on the RIKEN facility at ISIS, 1994.

Signing of the RIKEN agreement, 1990. The RIKEN-RAL Muon facility was constructed at ISIS by the Japanese research institution RIKEN. Left to right: Dr Paul Williams, Professor Oda (RIKEN), Bob Voss (ISIS), Professor Kanetada Nagamine (RIKEN), Sir William Mitchell (Chairman of the Science and Engineering Research Council).

Dr Colin Carlile, Instrument Scientist on IRIS installing a new detector bank. From 2001 to 2006 Dr Carlile was Director of the Institut Laue-Langevin.

The ISIS main control room.

Neil Grafton and John Ellis at the ISIS main control room, 1998.

The ISIS Experimental Hall, 2003. The hall is about the size of a football pitch.

Penchang Dai, University of Tennessee, preparing for his HET experiment, 2001.

The ISIS User Office, 2000.

Dr Steven Parnell setting up the spin exchange apparatus in RAL's new polarising filters laboratory where the technology was developed before being used on ISIS instruments, 2002.

Neil Meadowcroft inspecting the new main magnet power supply capacitor bank, 2002.

Professor Jon Goff and Dr Jose De Toro, Liverpool University, analysing their data on magnetic disorder in Fe/Mn superlattice, 2004.

Supriyo Ganguly of The Open University with a large sample on ENGIN-X. This is a prototype wing-box typical of that used on the Airbus 380, 2003. The ENGIN-X instrument is exclusively designed for making measurements on engineering components that can be up to a metre in size and weigh as much as 1.5 tonnes.

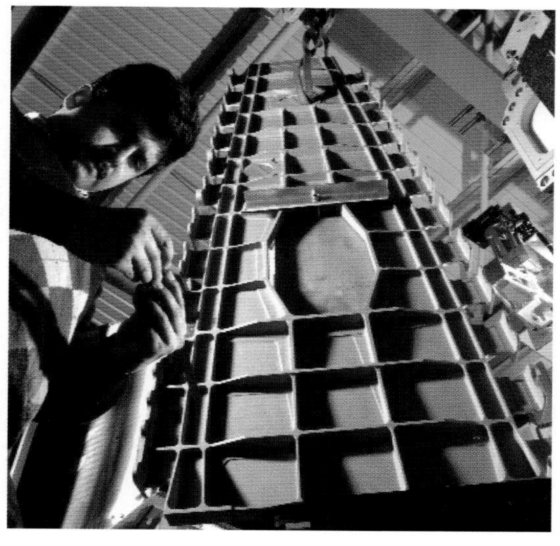

Alan Letchford looking down the inside of the vacuum vessel of the ISIS radio frequency quadrupole (RFQ) accelerator. This 4-rod RFQ was installed on ISIS as an upgrade for the existing but ageing Cockcroft-Walton pre-injector, 2003.

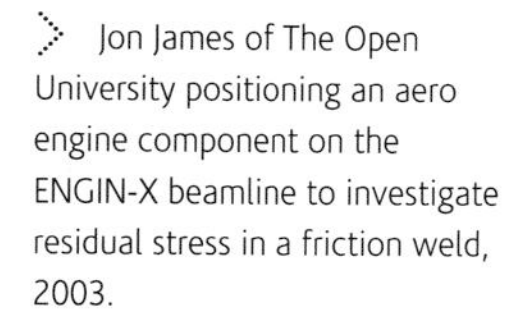

Jon James of The Open University positioning an aero engine component on the ENGIN-X beamline to investigate residual stress in a friction weld, 2003.

Peter Lock, University College London, preparing a sample of aqeous clay for an IRIS experiment, 2006.

Dr Chris Frost examines the detectors on the MAPS instrument, 2004. MAPS is ISIS's neutron scattering instrument designed to investigate single crystal samples.

Professor Sir David King opening the new ISIS Second Target Station technical support building during his visit to the facility for the 20th anniversary celebrations, 2004.

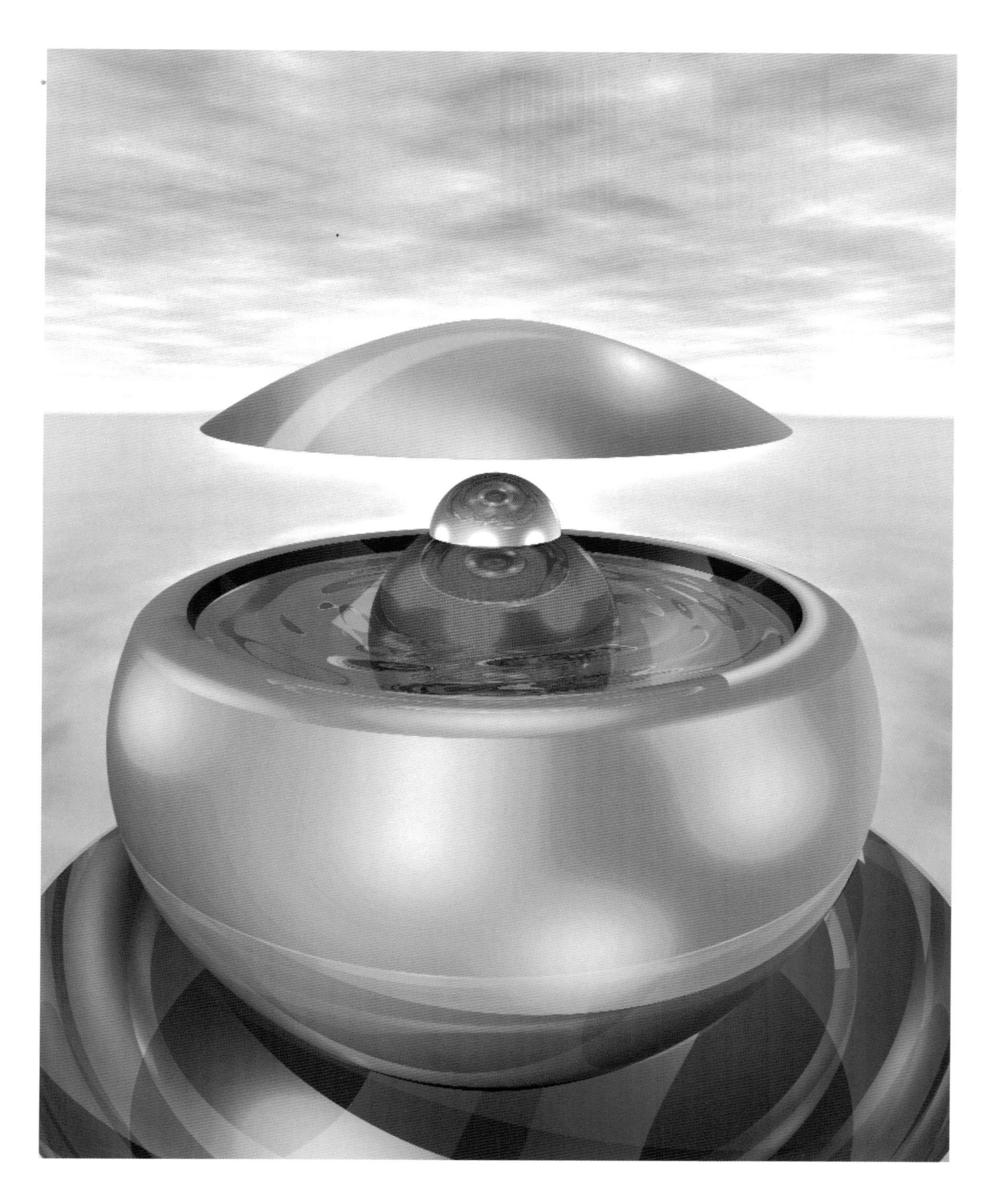

A hydroxide ion imposes structure on surrounding water molecules. Where there is water, there are negatively charged hydroxide ions which convert any oily material around them to soap. At least two distinct shells of water are found, forming a 'cup and saucer' arrangement, with the inner 'cup' containing about four water molecules. Over the top of the cup is a 'lid' of weakly bonded water molecules. As soon as a proton attempts to hop from a 'cup' water molecule onto the hydroxide ion, this molecule is replaced by one of the molecules from the 'lid'.

Credit: Created by Robert Dalgliesh with input from Kenneth Shankland and Adrian Hillier.

Dr Laura Cartechini from Italy's National Research Council holds an Etruscan eagle's head dating from c560 BC as Dr Winfried Kockelmann prepares the sample chamber on the ROTAX instrument at ISIS, 2003.

Whilst most people associate the term 'liquid crystal' with clock displays and watches, there are many other types of liquid crystal that we encounter every day. They are found, for example, in shampoos and laundry detergents, where the active molecules (called surfactants) assemble into a rich variety of aggregate structures such as spheres, discs or, as shown here, long tubes. These aggregates, called micelles, may range in size from a few nanometres to several thousand nanometres. Even the micelles themselves can order into regular arrays, just like atoms in normal crystals. The technique of small-angle neutron scattering is used to examine this ordering and probe the size and shape of the micelles and this has led to improvements in the formulation and manufacturing of a number of surfactant-based products in common use.

Credit: Created by Robert Dalgliesh with input from Kenneth Shankland and Adrian Hillier.

Professor John Wood and Dr Andrew Taylor opening the mound, 2006.

ISIS is doubling its capacity and upgrading its capability by constructing a Second Target Station.

After moving 300,000 m^3 of chalk, building began in winter 2005. With the steel infrastructure complete, the facility begain installing its first seven instruments in June 2007 and is on target to start an experimental programme in 2008. A whole new capability, optimised for studies in the key science areas of soft matter, biomolecular systems and advanced materials – nanoscale science – is fast becoming a reality.

Dr Helen Jarvie, Centre for Ecology and Hydrology and Dr Stephen King preparing samples for an experiment on water pollution. The experiment successfully used neutron scattering techniques previously used on a wide range of materials such as ceramics and alloys, to samples from a fresh water environment for the first time.

He Wei, from the Chinese Academy of Science, testing components for the Chinese Ion Source, 2007.

ISIS supports an international community of around 1600 scientists who use neutrons and muons for research in physics, chemistry, materials science, geology, engineering and biology.

One of the many visiting research scientists using ISIS, 1996.

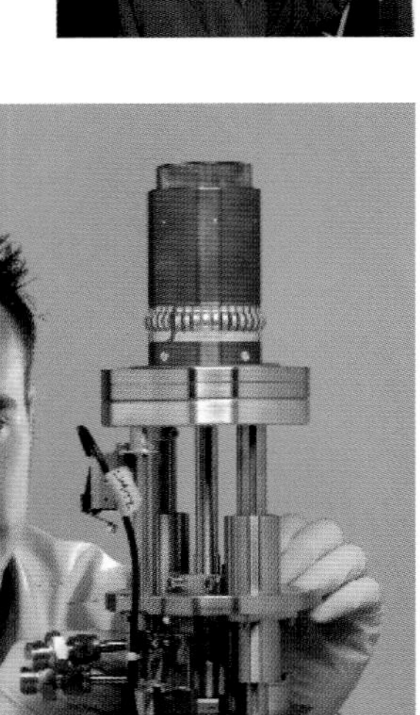

Chris Benson from the ISIS Project Engineering group working on the tuner for the ISIS radio-frequency quadrupole accelerator, 2004.

ISIS

Lasers

Ultra High Vacuum Chamber in the Central Laser Facility. A view from the end of the line-of-sight mass spectrometer looking through one of the chamber's windows, 2005.

For over 30 years, the Central Laser Facility (CLF) has been a world leader in the provision and application of ultra-fast and high intensity lasers to a broad, international user community.

The CLF was established in the mid 1970s, following the recommendation of a panel of leading academics that recognised the benefits of developing a national facility in the rapidly growing field of high power laser science.

Work at the CLF has been pioneering since its inception, both in terms of the capabilities of the lasers and in the science performed on them. From single beam interaction studies in the late 1970s, the facility's technological capabilities have diversified from photons to GeV electrons, MeV protons, X-ray lasers, giga-gauss magnetic fields, and gamma rays. In 2004, the CLF was proud to be awarded a Guinness World Record for Vulcan as "the highest intensity focused laser in the world".

The CLF has successfully remained at the cutting-edge of research, through coupling the unrivalled knowledge and expertise of its staff and a programme of continual development with the scientific programmes of its international user community, ensuring that their demanding requirements are met. The facility now supports a vast range of applications in energy, bioscience, material science, environment and accelerator technology, and other fields. Many emergent innovations are already benefiting society at large, for example, applications of Raman spectroscopy to counterfeit drug detection, screening for liquid explosives and non-invasive medical diagnosis.

With the completion of the recent Astra Gemini upgrade, and an exciting range of programmes, further upgrades and new facilities on the horizon, the next 30 years at the CLF promise to be as successful and inspiring as the last.

Dr Paul Williams at the Central Laser Facility (CLF) inauguration, 1977.

The CLF inauguration party, 20 June 1977. Left to right: Professor Alan Gibson (first Head of the Central Laser Facility), Professor Dan Bradley (leading academic), Sir Sam Edwards (Science Research Council Chairman) and Dr Godfrey Stafford (Director of Rutherford).

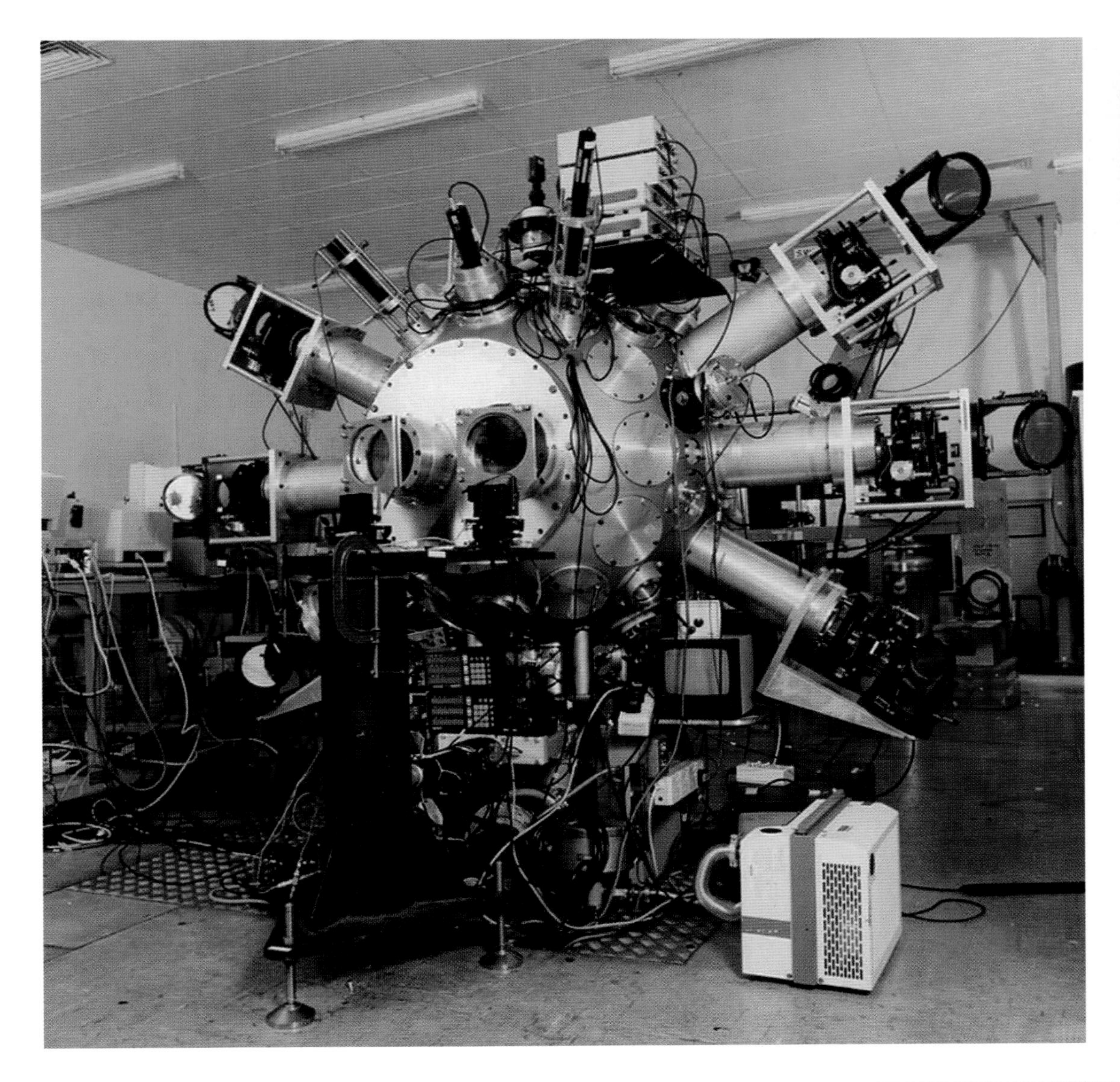

A novel 6 beam line focus target chamber enabled unique X-ray laser work. 1986.

Professor Alan Gibson retired in 1983 after exemplary leadership for the first seven years of the CLF.

The SPRITE KrF laser, the most powerful laser of its type, was commissioned in 1982. Left to right: Dr Graeme Hirst, Jim Lister, Chris Ansah, Dr Mick Shaw, Chris Hooker, Dr Erol Harvey. Graeme, then a post-graduate researcher, is now responsible for 'Laser-Accelerator Applications' at RAL.

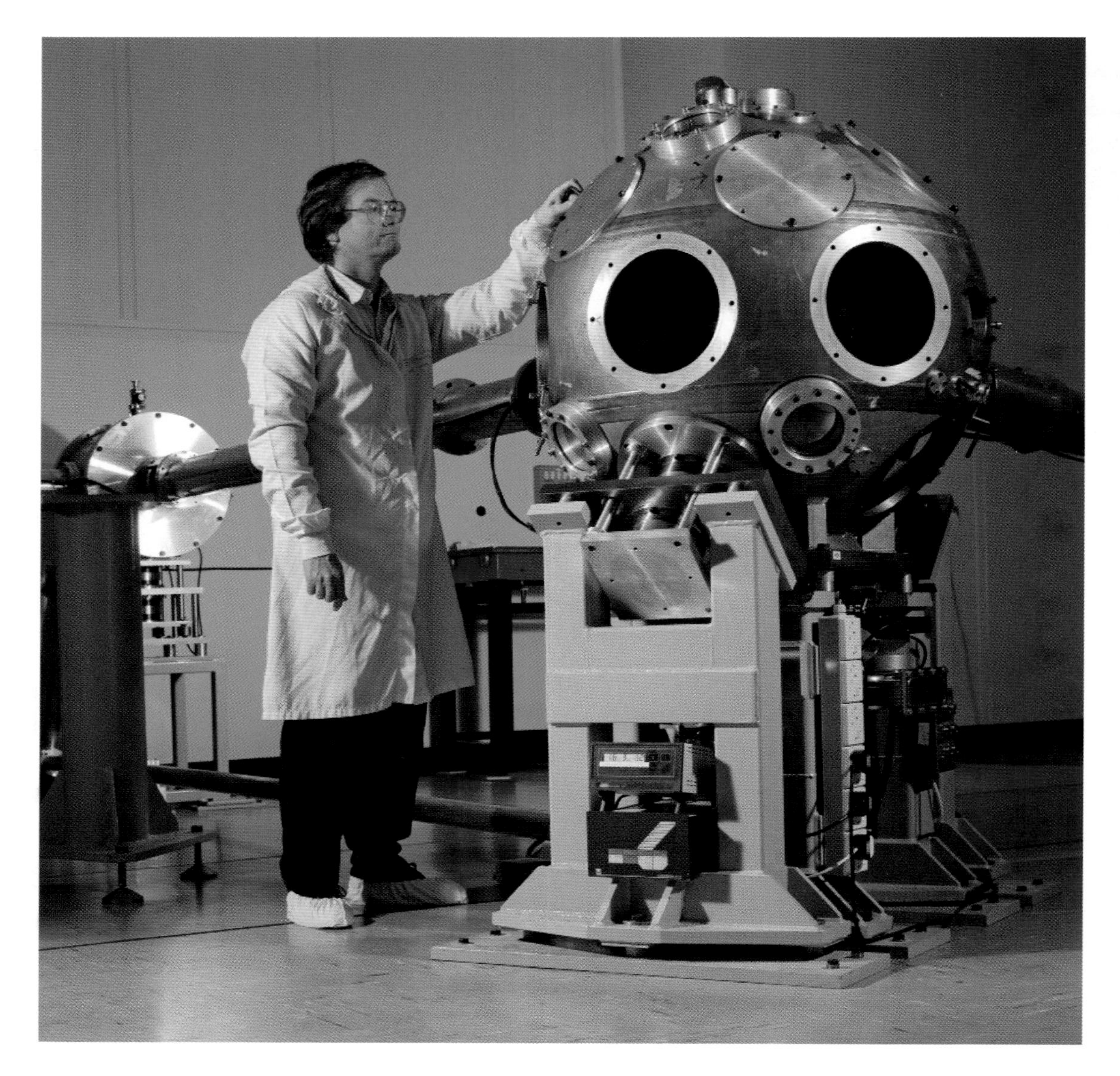

The SPRITE KrF laser was upgraded, resulting in the comissioning of Titania in 1996, enabling high repetition rate, high efficiency gas laser facilities.

Professor Peter Norreys in the Titania laser target area, 1996.

Dr Paul Williams, Director RAL, being interviewed at the Titania laser inauguration, Dr Graeme Hirst looking on, 1996.

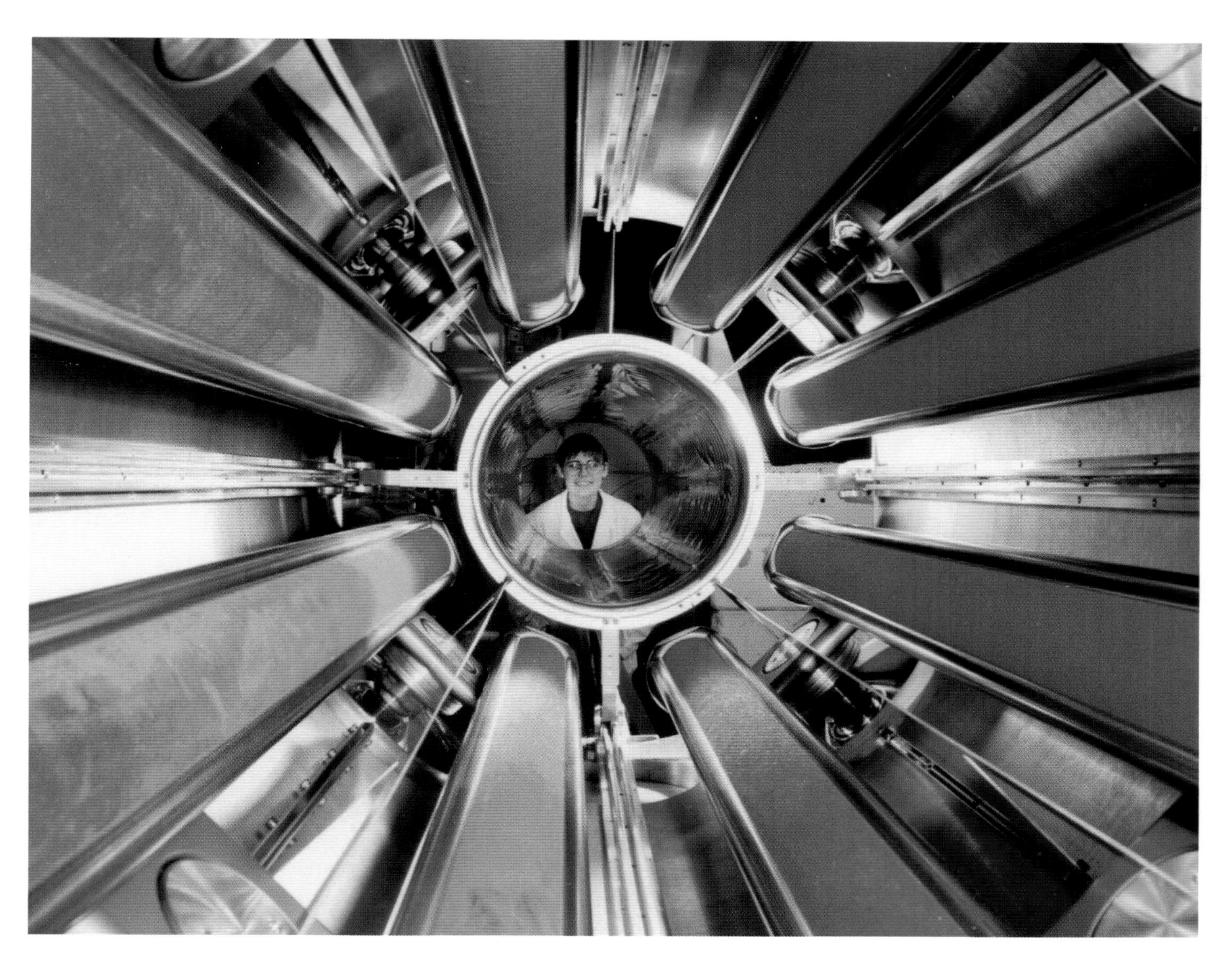

Apprentice Sally-Ann Brind inspects the laser chamber, 1994.

Aligning mirrors along the path of the laser.

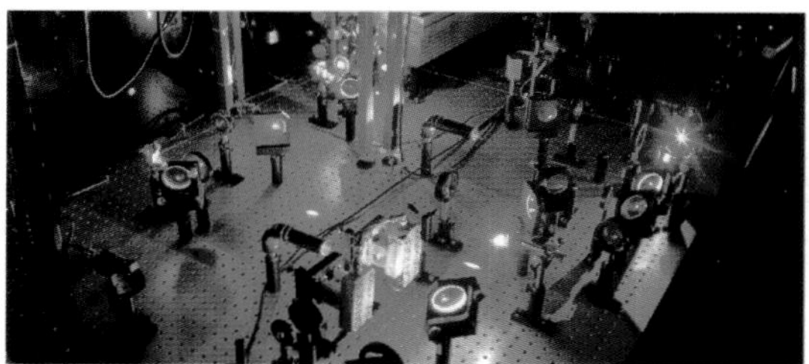

Disk amplifiers in Vulcan, 1998. Colin Danson (front) and Bob Bann.

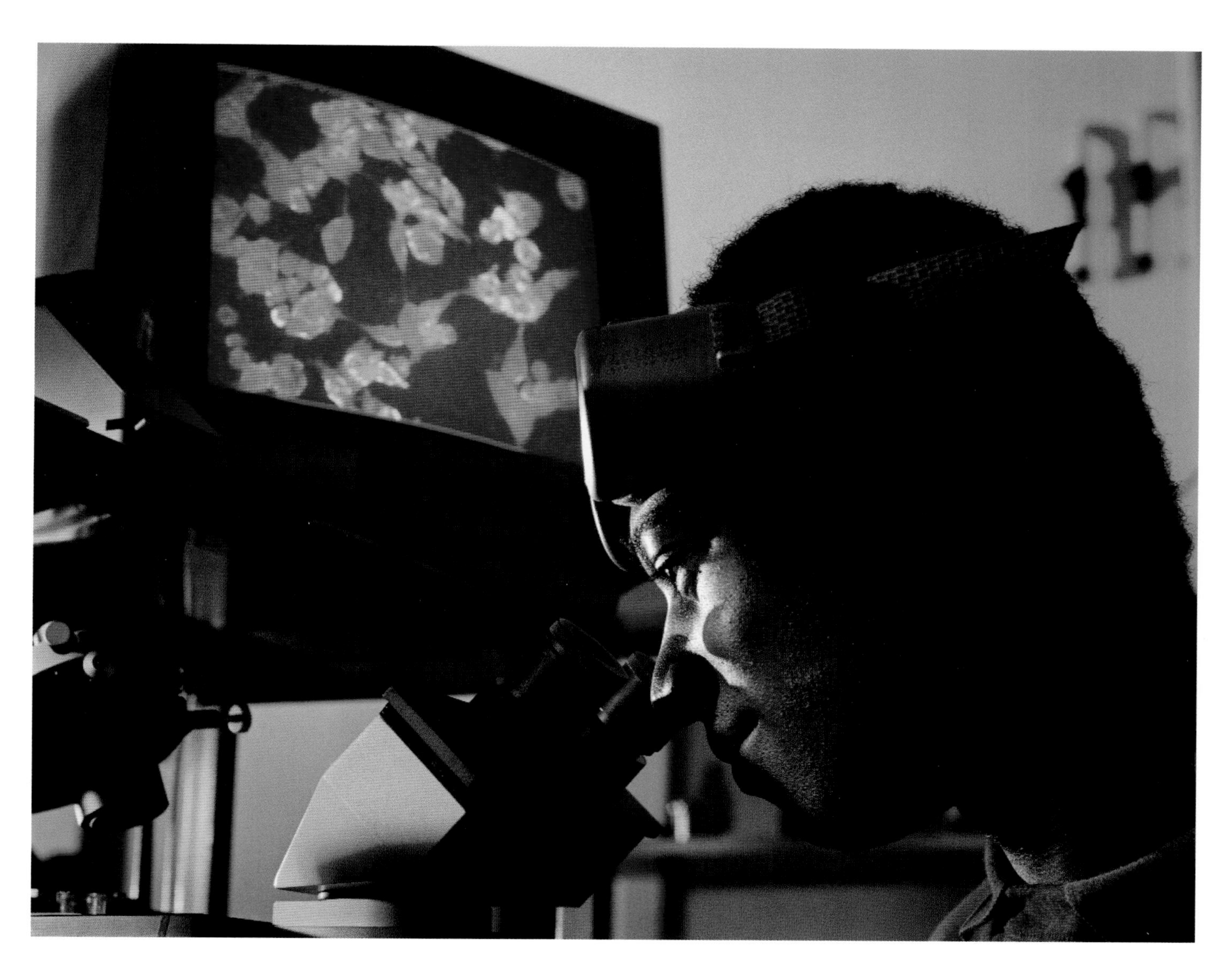

Dr Stan Botchway using the Confocal Microscope in the Laser for Science Facility, just one of a range of services available to users, 2000.

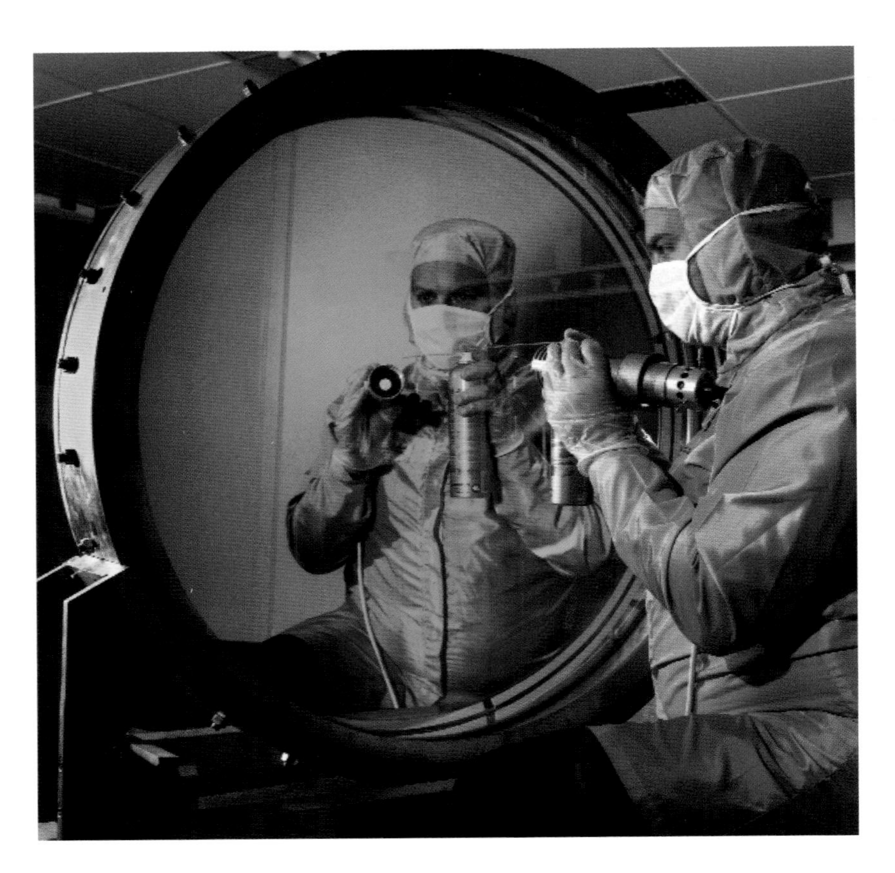

Trevor Winstone tests a large mirror for the Vulcan Petawatt Facility, 2000.

The interaction chamber for the Vulcan Petawatt Facility upgrade being carefully manoeuvred into the purpose-built target area building, 2000.

The first LASERLAB-EUROPE User Meeting, hosted by St. Catherine's College, Oxford, was organised by RAL in 2005. A visit to the laboratory proved a popular component of the meeting.

In 2004, two world records were awarded to the Rutherford Appleton Laboratory. It was official – Vulcan was named the world's highest intensity focussed laser whilst ISIS was named the world's most powerful pulsed neutron spallation source. Ian Gardner (ISIS) and Colin Danson (CLF) received certificates from Guiness World Records' David Hawksett (centre) on behalf of the facilities.

RAL

A student examining an optic during the CLF Training Week, 2005. The CLF hosts an annual week long course, involving classes and practical work, to provide training for PhD students in using the Vulcan high power laser.

Chris Gregory aligning mirrors during the CLF Training Week, 2005.

Quanli Dong, University of York, investigating a laser diffraction grating, 2003.

A team from Imperial College, London in the Astra laser target area, 2004. Left to right: Alec Thomas, Chris Murphy and Stuart Mangles.

Visitors inspecting a crystal amplifier during a public open afternoon, 2004.

Visitors exploring the Astra laser target area, March 2004.

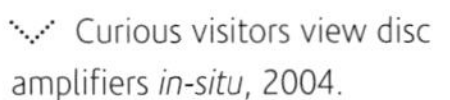
Curious visitors view disc amplifiers *in-situ*, 2004.

Rob Clarke guiding visitors from the Royal Society for the Encouragement of Arts, Manufactures and Commerce, around the Vulcan Petawatt Facility, 2007.

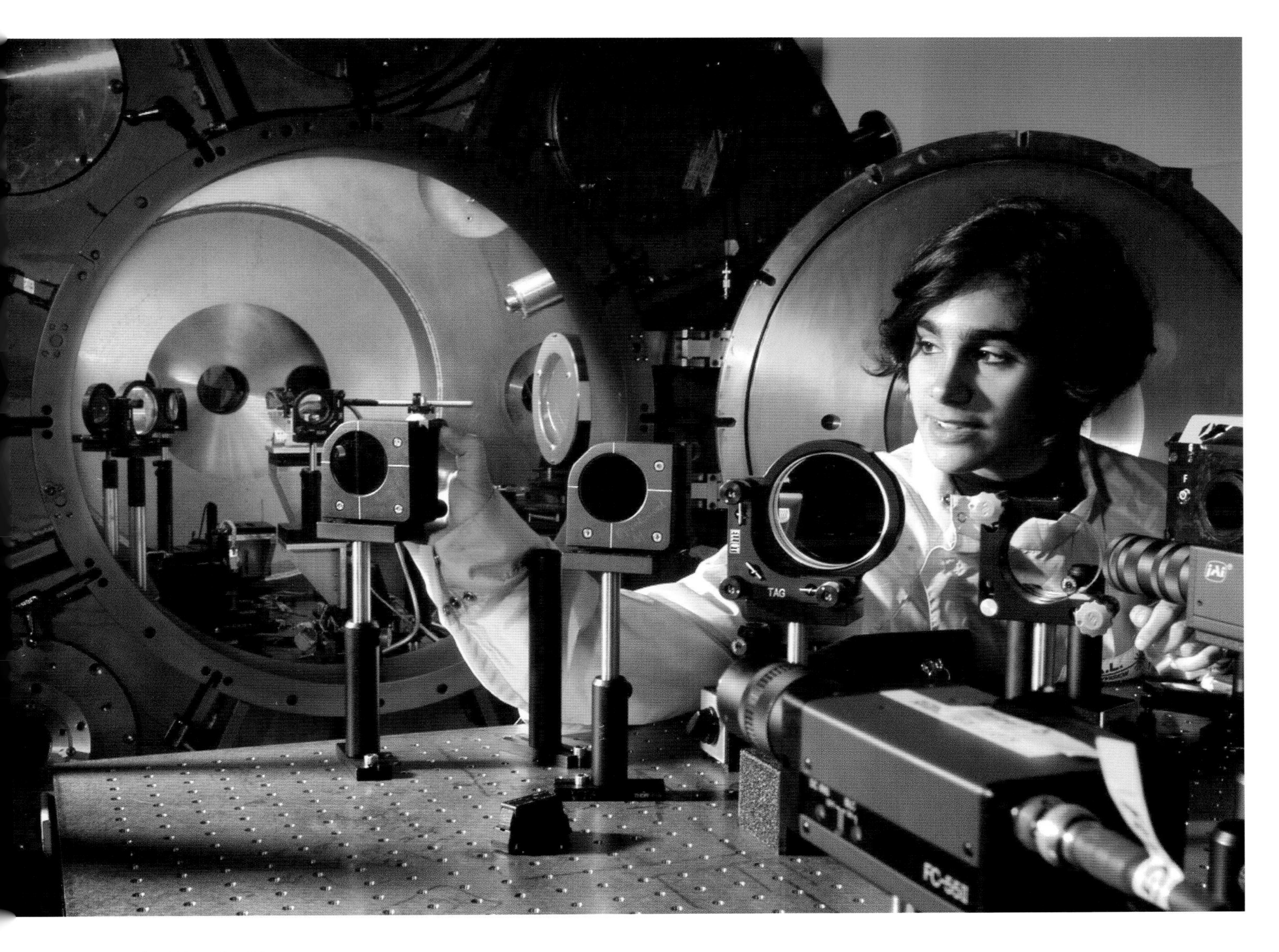

Vulcan's Target Area West. Alessandro Rovasio (Laboratoire pour l'Utilisation des Lasers Intenses, France) preparing apparatus for a laser-induced shock waves experiment, 2005.

Robert Kraty, a sandwich course student, inside the target chamber of Target Area East, placing a target into the set-up for an X-ray laser experiment, 2001.

Some targets made at RAL are almost small enough to fit through the eye of a needle.

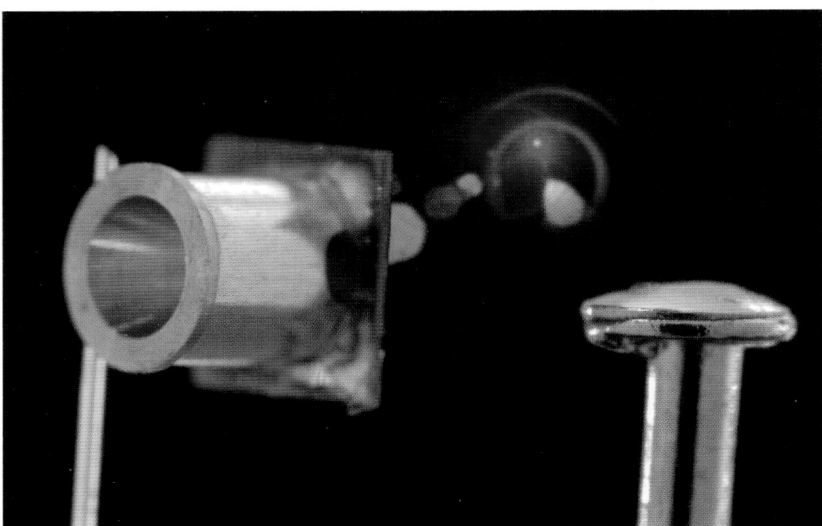

A tiny target (shown next to a pin head) prepared by specialist technicians in the CLF, 2005.

Ali Mohammed (left) and Dr Andy Kidd (right) analysing an experiment using the CLF diagnostic kit, 2005.

Chinese collaboration on a tour of the CLF with new facility Director Professor Mike Dunne, 2006.

Dr Pavel Matousek and colleague Dr Charlotte Eliasson have developed and tested a technique to verify drugs without removing them from their packaging. The technique known as Spatially Offset Raman Spectroscopy, SORS, was developed at RAL, 2007.

Oleg Chekhlov working in the Astra Gemini laser area preparing it for the first firing. Image shows Oleg inspecting optics inside one of the compressor chambers, 2007.

Bryn Parry working in the Astra Gemini laser area fine tuning optics on an amplifier table, 2007.

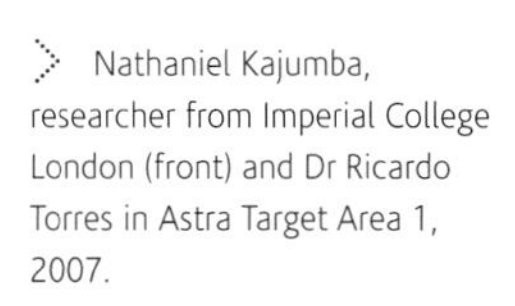

Nathaniel Kajumba, researcher from Imperial College London (front) and Dr Ricardo Torres in Astra Target Area 1, 2007.

The next generation of scientists and engineers are inspired through visits and work experience placements at the laboratory. Here students explore the Astra facility, 2005.

Space

An image taken with NASA's Hubble Space Telescope showing bright, blue, newly formed stars in the centre of a star-forming region in the Small Magellanic Cloud. At the heart of the star-forming region lies star cluster NGC 602.

Credit: NASA, ESA, and the Hubble Heritage Team (STScI/AURA) - ESA/Hubble Collaboration

The Rutherford Appleton Laboratory has played a rich and unique role in the history of space science over the last 50 years.

The modern day Space Science and Technology Department had its roots at Ditton Park in Slough as the Radio Research Station, later the Radio and Space Research Station and then the Appleton Laboratory. Ditton Park became a major centre for UK space research when the launch of Sputnik-1 provided its scientists with the opportunity for a quickly contrived experiment using a cathode-ray detector. This marked the start of a history closely linked with that of space science. The year was 1957, the International Geophysical Year – the year seen as marking the start of mankind's journey into space. In the years that followed, the growth of space science activities meant that the support of the UK space programme became the main focus of the Appleton Laboratory.

In 1979, the Appleton and Rutherford Laboratories merged, and the space programme moved to the Rutherford Appleton's Chilton site and the Chilbolton Observatory in Hampshire. Since then, the Laboratory's field of space research has broadened greatly. The skyline at Chilton was dominated by the 12-metre S-band antenna used for tracking, commanding and receiving data from NASA's Infra-Red Astronomy Satellite (IRAS) which was successfully launched in 1983. At the time IRAS was NASA's largest ever space science mission and all operations were undertaken at RAL.

Today, the Space Science and Technology Department (SSTD) provides world leading science and technology development, space test facilities, instrument and mission design, and data curation and processing systems for customers around the world. SSTD is now a premier European space department and has been involved in over 150 missions in recent years including the groundbreaking SOHO and STEREO solar missions, the Earth Remote Sensing missions ERS-1, ERS-2 and ENVISAT, and solar system missions such as the Rosetta cometary lander, the Cassini/Huygens mission to Saturn and its moon Titan, and continuing work on MIRI for the James Webb Space Telescope.

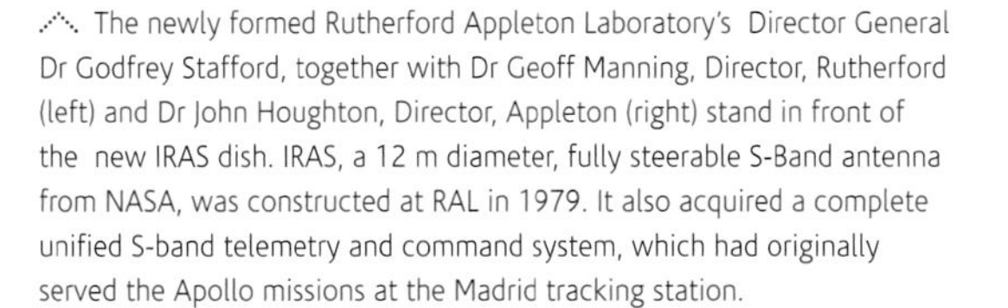

The newly formed Rutherford Appleton Laboratory's Director General Dr Godfrey Stafford, together with Dr Geoff Manning, Director, Rutherford (left) and Dr John Houghton, Director, Appleton (right) stand in front of the new IRAS dish. IRAS, a 12 m diameter, fully steerable S-Band antenna from NASA, was constructed at RAL in 1979. It also acquired a complete unified S-band telemetry and command system, which had originally served the Apollo missions at the Madrid tracking station.

The Radio Research Station (later named the Appleton Laboratory after Sir Edward Appleton) was founded in 1920 at Ditton Park, Slough. The Appleton Laboratory merged with the Rutherford Laboratory in 1979 to form the Rutherford Appleton Laboratory (RAL).

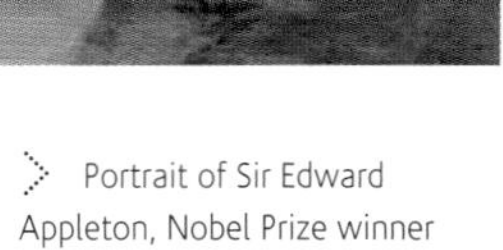

Portrait of Sir Edward Appleton, Nobel Prize winner and discoverer of the ionosphere.

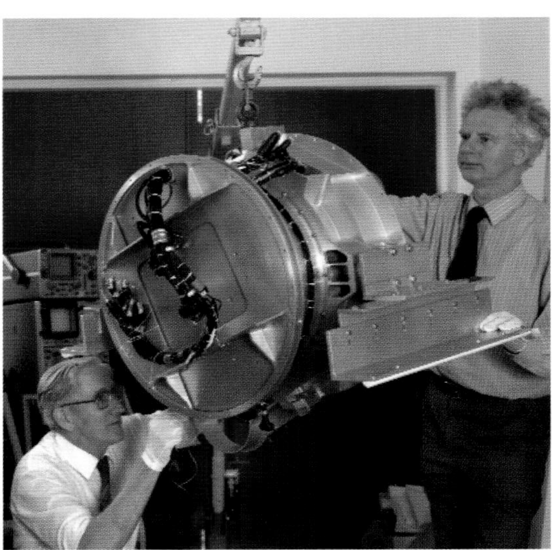

Researchers from Birmingham University's Space Science Department inspecting ROSAT Wide Field Camera before loading into the simulated space environment facility, 1986. RAL was responsible for overall project management, design and manufacture of the camera's filter system. ROSAT, an X-ray observatory developed through a cooperative programme between Germany, the US, and the UK, was launched on 1 June 1990. The expected mission lifetime was two years, the final operational lifetime exceeded this by six years.

AMPTE spacecraft in final assembly, 1984. AMPTE (Active Magnetospheric Particle Tracer Explorers) explored plasma phenomena in near-Earth space. The AMPTE mission was comprised of three spacecraft which explored the magnetosphere from 1984 to 1986. The wealth of data collected is still valuable today as they are still among the best available data for studies of many magnetospheric phenomena.

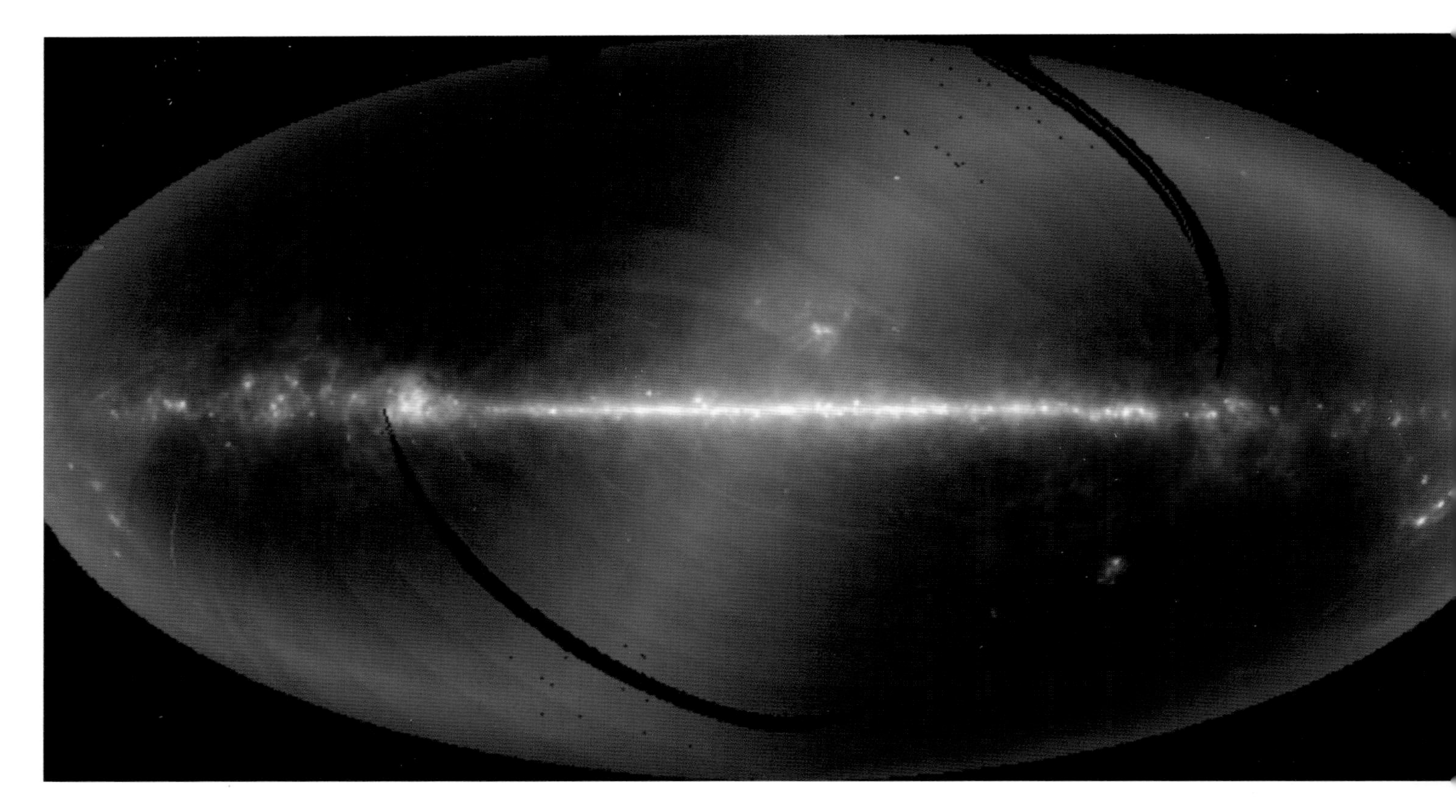

Nearly the entire sky, as seen in infrared wavelengths and projected at one-half degree resolution, is shown in this image, assembled from six months of data from the Infrared Astronomical Satellite (IRAS), a joint project of the US, UK (RAL was responsible for daily operations) and the Netherlands. The bright horizontal band is the plane of the Milky Way, with the middle of the Galaxy located centrally. Launched in January 1983, IRAS ended its mission ten months later. The mission increased the number of catalogued astronomical sources by about 70%, detecting approximately 350,000 infrared sources. Discoveries included a disk of dust grains around the star Vega, six new comets, and very strong infrared emission from interacting galaxies as well as wisps of warm dust called infrared cirrus which could be found in almost every direction of space. IRAS also revealed the Milky Way.

Credit: NASA/JPL-Caltech

The artist's rendering shows IRAS in its 560-mile-high, near-polar orbit above the Earth. From this vantage point, IRAS searched the sky for stars and other infrared-emitting sources, unhampered by the obscuring effects of Earth's atmosphere.

The Chilton ionosonde located at RAL, 1996. This instrument is a Lowell Digisonde, DPS1 and it continues the sequence of soundings started in Slough in 1931. In order to ensure that the change of site did not affect the data sequence, operation of both the Chilton and Slough ionosondes was carried out for over a year.

An eclipse is the nearest nature comes to turning off the Sun. When this happens, the temperature drops, changing the wind pattern as air contracts in on the eclipse region. In the absence of the ionising radiation from the Sun, our ionosphere rapidly decays, affecting the propagation of radio signals. The shadow of the moon races through the atmosphere at supersonic speeds, causing waves to spread through the atmosphere from the eclipse region. A collaboration of scientists from universities and RAL used the 1999 eclipse to study these theories further.
Credit: Paul Andrews

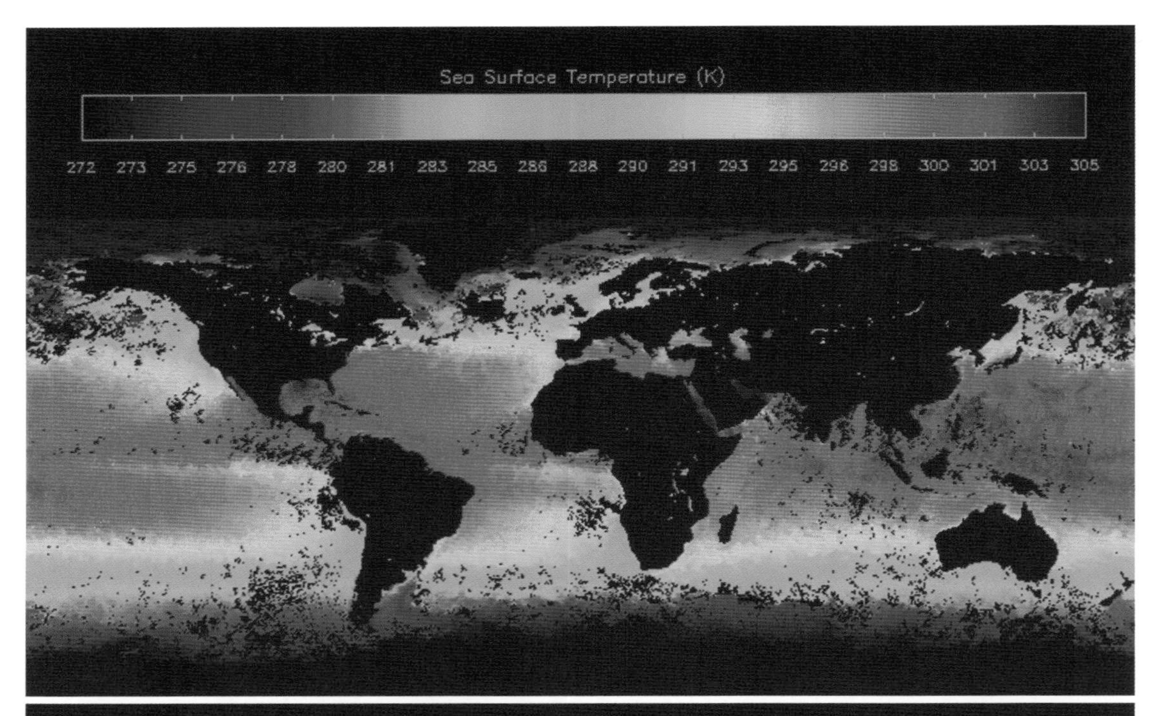

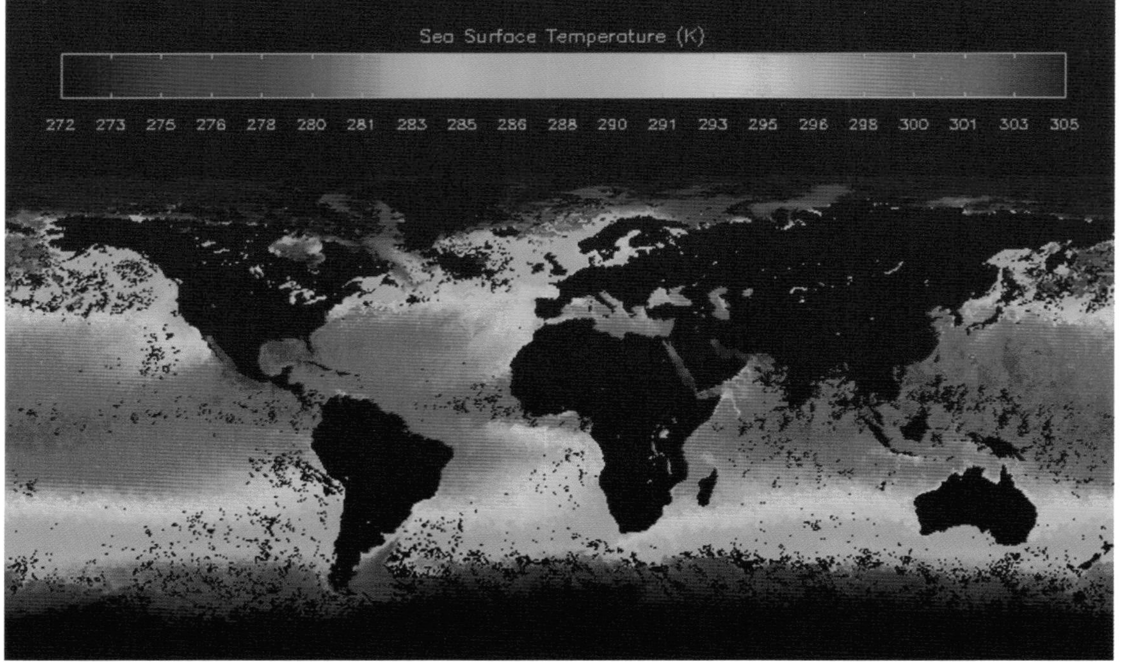

Sea surface temperatures as measured by the Along Track Scanning Radiometer-1 (ATSR) built at RAL. Temperatures in a normal year (July 1995 – top) compared to those during an El Nino (July 1997).

Helen Mapson-Menard from the University of Birmingham with the SOLAR-B mass thermal model in the environment testing tank at RAL, 2002.

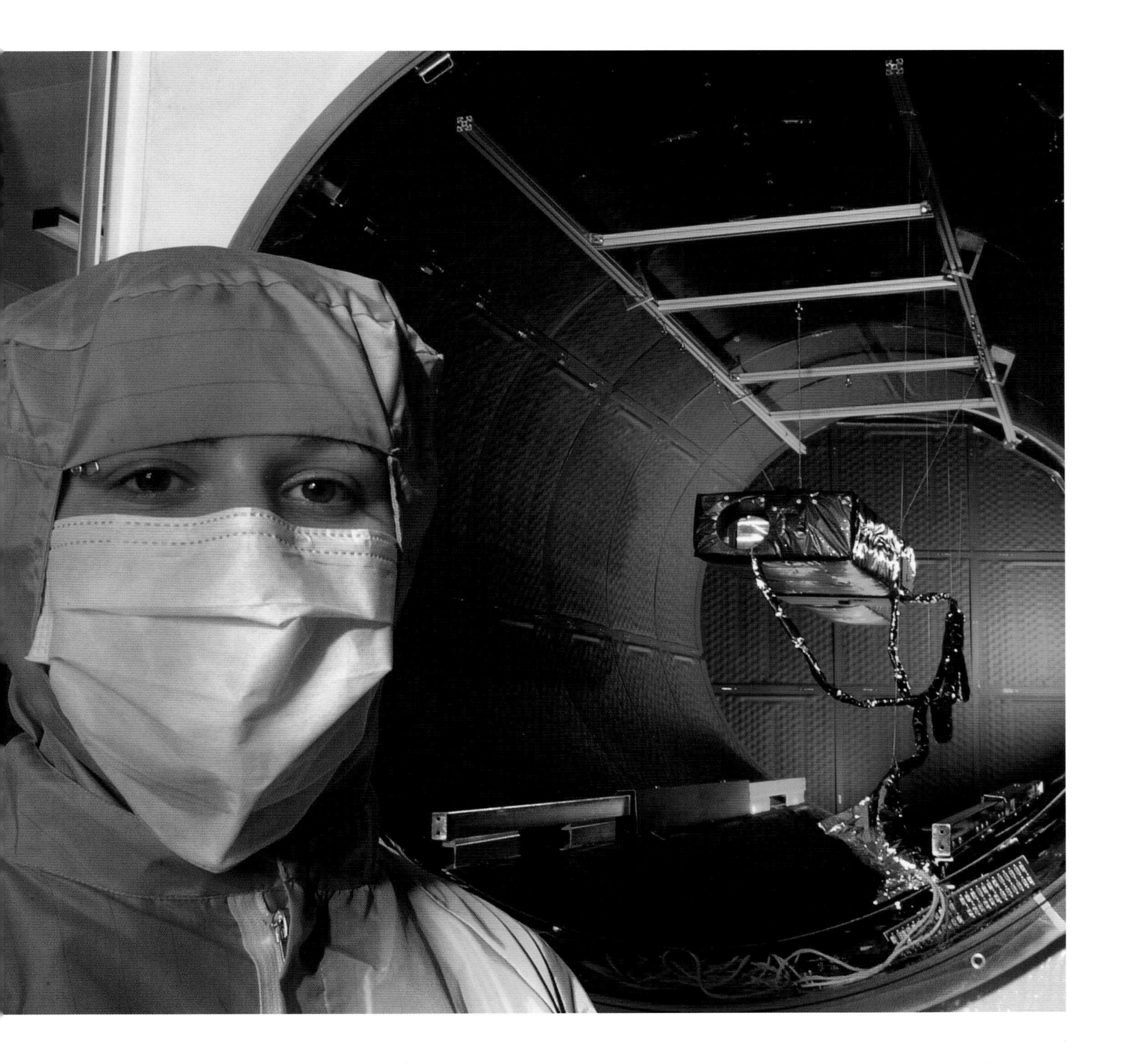

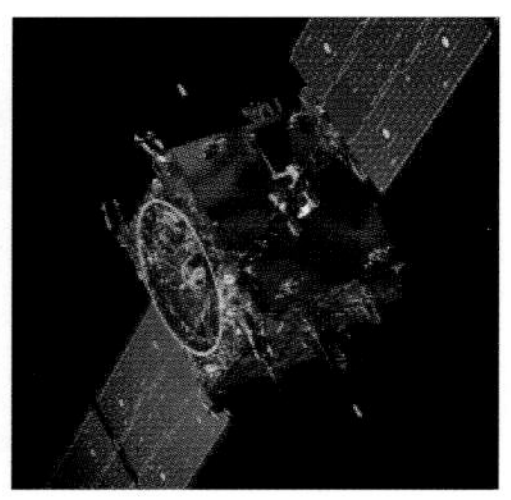

Artist's impression of Smart-1. Launched on 27 September 2003, Smart-1 successfully crashed into the Moon on 3 September 2006 at 05:42:21.759, bringing to an end a very eventful and challenging mission.
Credit: ESA

Dr Sarah Dunkin studying maps of the Moon's surface taken by the Demonstration Compact Imaging X-ray Spectrometer (D-CIXS) on Smart-1, 2003. RAL lead the D-CIXS team (for the European Space Agency) for the entire project lifecycle. Smart-1 utilised the miniaturisation of instruments for the first time.

Beagle 2 being prepared for testing in the Space Test Chamber. Beagle was part of the mission Mars Express. Despite the much publicised failure of Beagle 2's landing, much of the underlying technology now has wide-ranging applications.

Local school children learning about Mars (the Red Planet) at a schools' activity day, 2002.

Mars Express (the Spacecraft) is performing exceptionally well. Artist's impression.
Credit: ESA

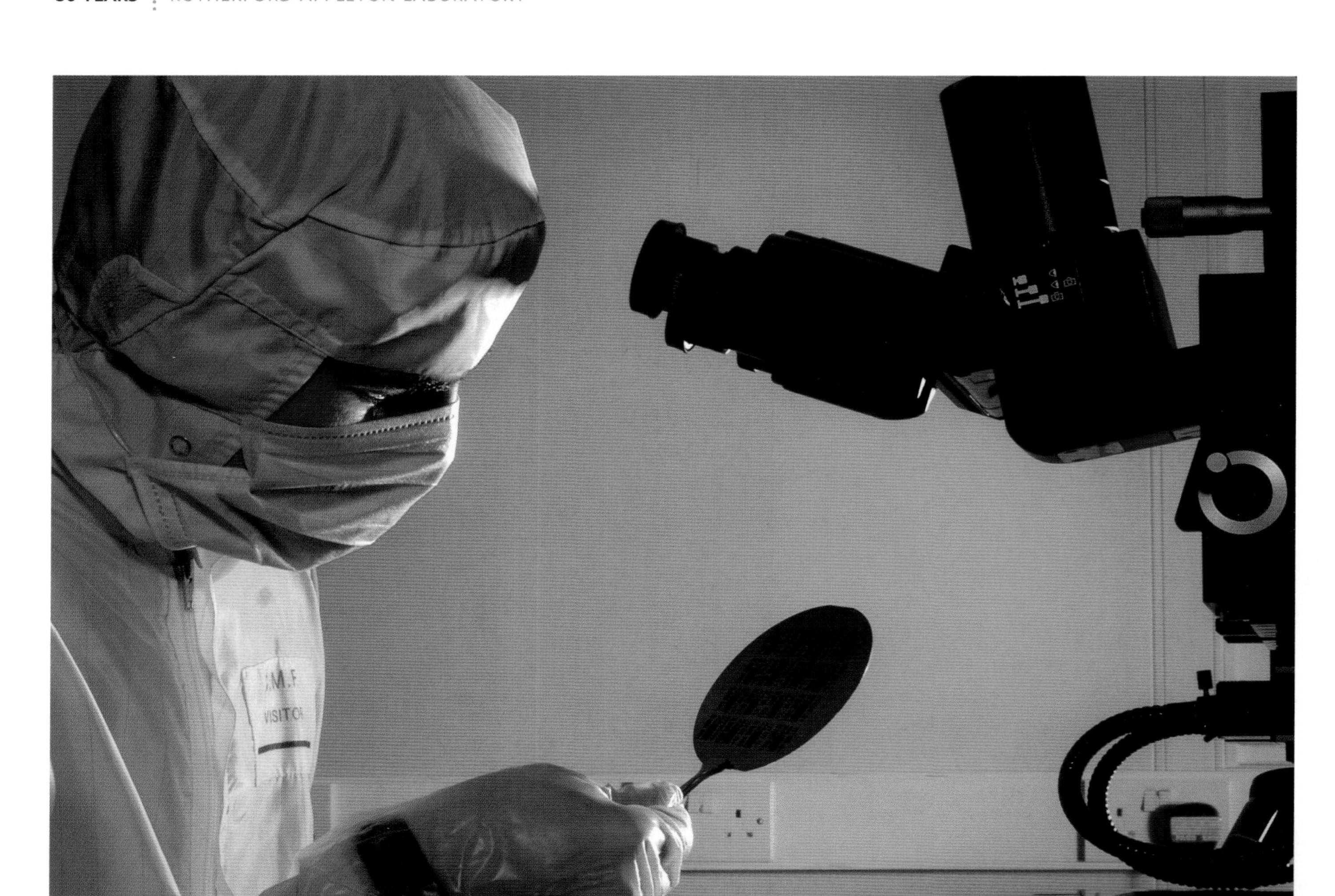

pA key development in terahertz technology was achieved through the Star Tiger project, a collaboration with ESA. The project team, which comprised 11 scientists from across Europe, produced the world's first compact terahertz imager capable of working in two frequencies, 2002.

Jonathan James demonstrating a potential security application of the terahertz imager, 2003. The technology was developed further by the spin-out company ThruVision. ThruVision's products have been successfully trialled at national airports.

Mike Trower (left) and Eric Clarke (right) complete testing of the Geostationary Earth Radiation Budget (GERB1) instrument, 2004. Carried on board the Meteosat Second Generation satellite which was launched in 2005, GERB1 is measuring the Earth radiation budget.

Transit of Venus, 2004.

Pat Trafford and Matt Burton watching the transit of Venus through solar glasses.

Dr Paul Soler and son watching the transit of Venus safely via a reflection of the Sun projected onto a piece of paper, 2004.

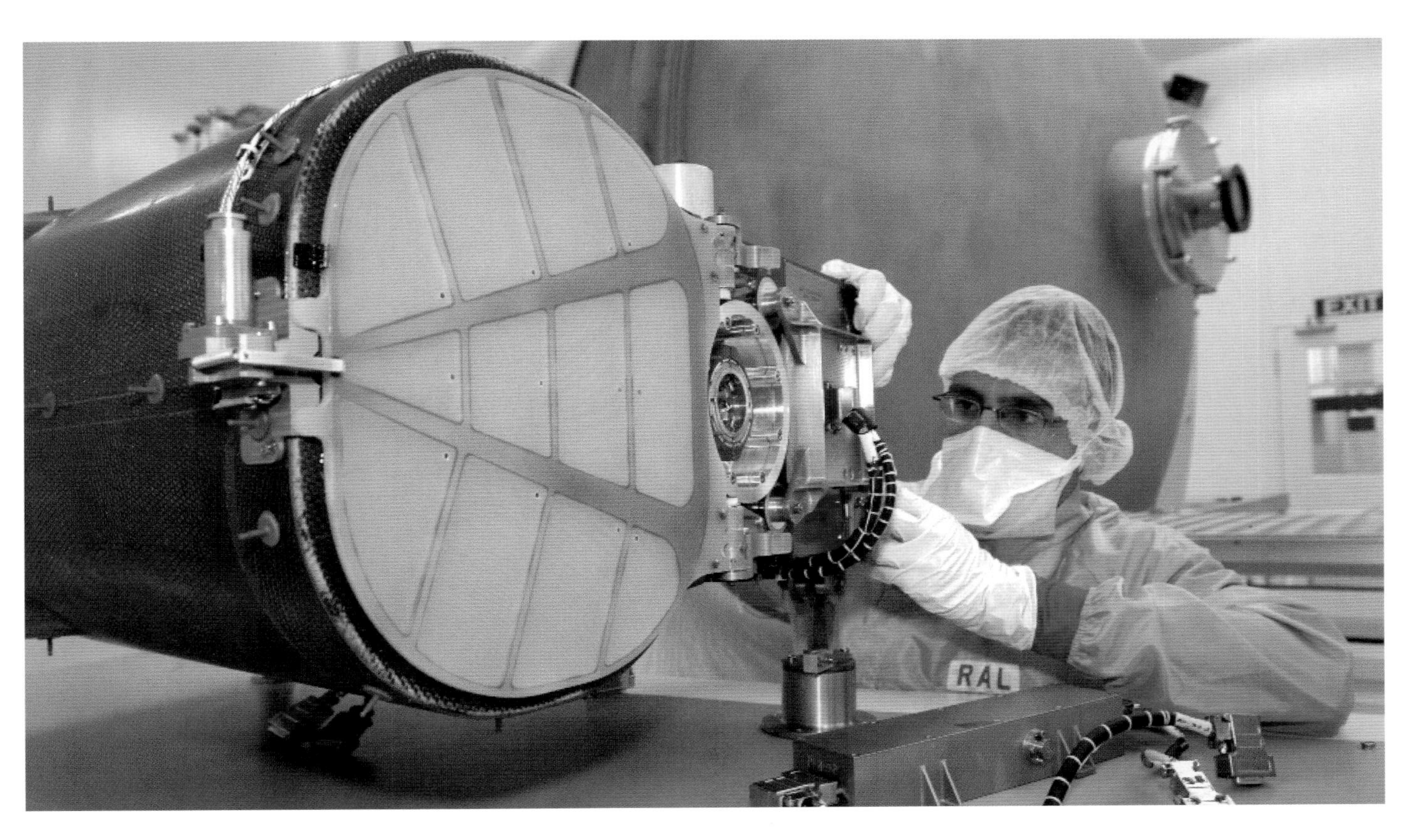

RAL developed a high-resolution optical camera for the TopSat programme, producing 2.5 m resolution panchromatic images and 5 m resolution colour images (shown under construction in 2004). Through the spin-out company Orbital Optics, which is designing and building low cost, high resolution cameras for use in space, the UK is leading the way in cost efficient imagery.

Nigel Cross, National Physics Laboratory (left) and Paul Greenway (right) at the TopSat launch event, 27 October 2005.

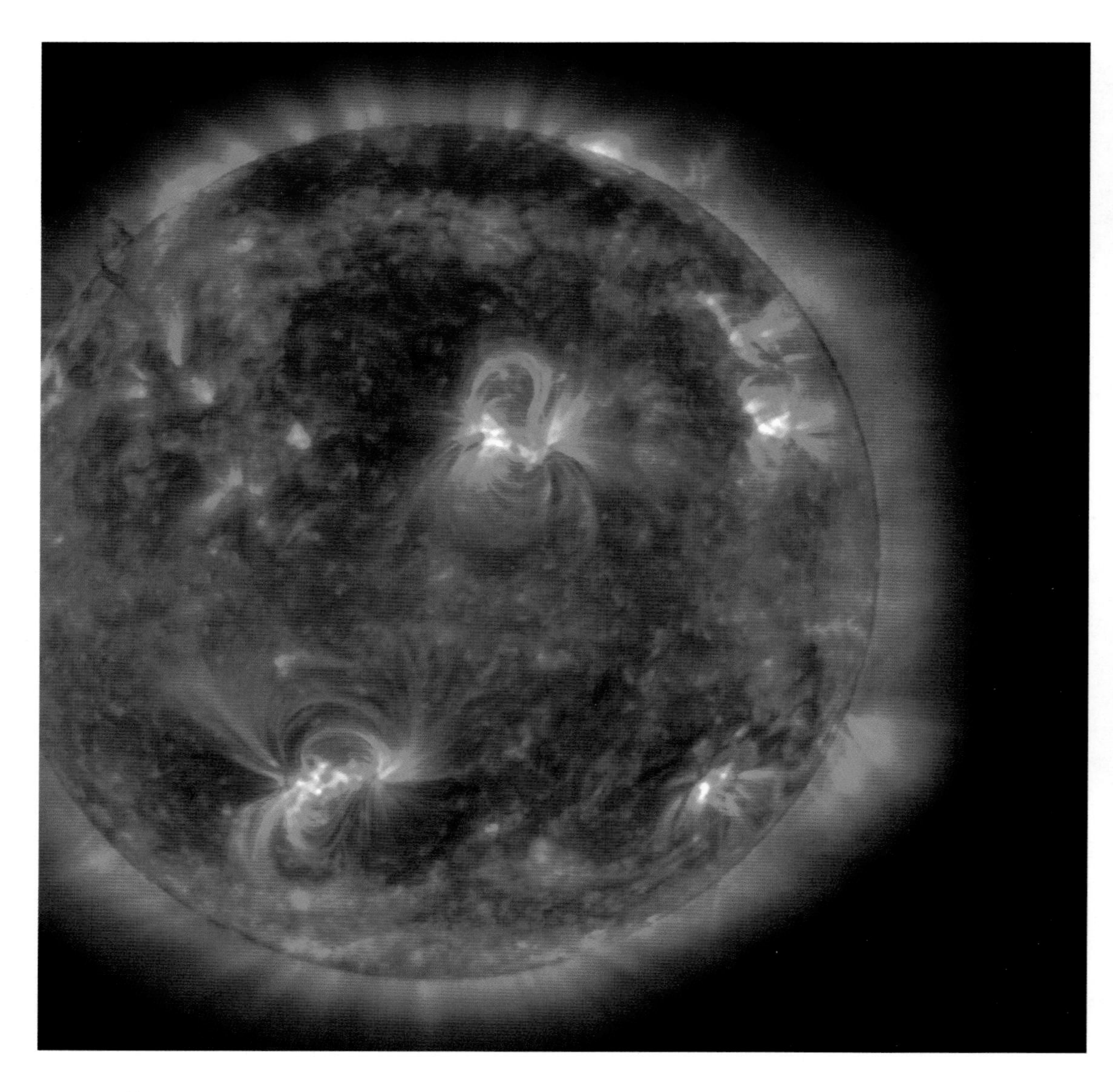

Image of the Sun captured by SOHO showing the solar corona at a temperature of about 1.3 million K and numerous solar loops. SOHO is studying the Sun, from its deep interior, through its atmosphere, out to the heliosphere, including the solar wind and its interaction with the interstellar breeze.

Professor Richard Harrison, Head of Space Physics, presenting HRH Prince Michael of Kent with an image of SOHO during his visit in 2004.

SOHO's 10th Anniversary in 2005. Professor Len Culhane, then Head of the Mullard Space Science Laboratory (left) and Dr Alan Gabriel, former Division leader in RAL's Space Science and Technology. Alan left RAL in 1986 to become director of the Institut d'Astrophysique Spatial in Orsay near Paris but is still very active in solar physics research, collaborating with the RAL solar group.

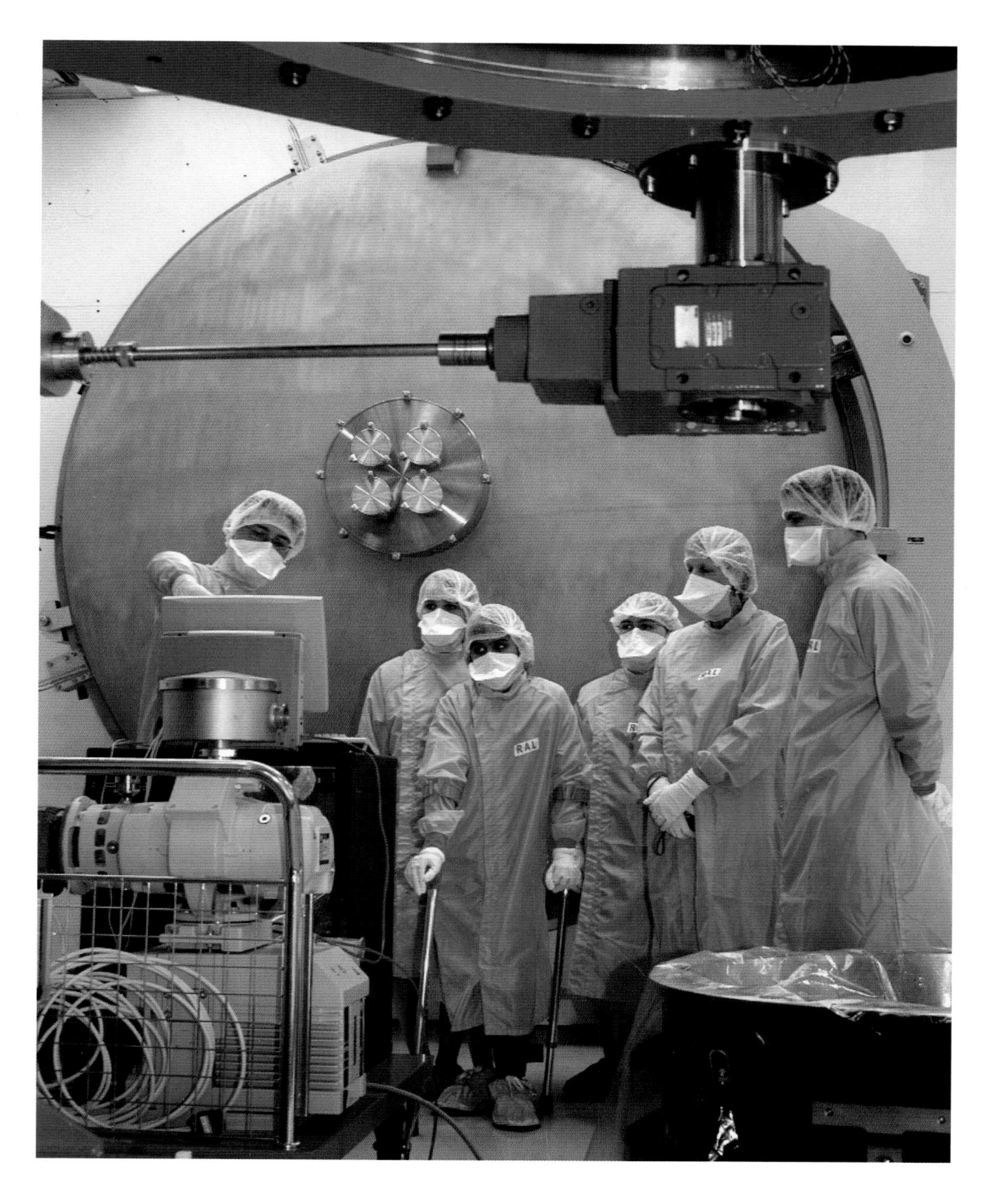

Research Councils UK's 'Science Race' competition winners visiting RAL as part of their prize, 2005.

Due for launch in 2013, the James Webb Space Telescope (JWST) will carry a set of four sophisticated instruments to enable superb imaging at visible and infrared wavelengths, together with spectroscopic modes to investigate the chemistry and evolution of the objects populating our Universe. One component, the Mid Infrared Instrument (MIRI), will enable the properties of materials forming around new born stars to be studied in unprecedented detail, and coronagraphs will allow direct imaging of massive planets orbiting other stars. Image shows Sam Heys and Paul Eccleston checking the MIRI satellite STM instrument in the Space Test Chamber, 1996.

The JWST is an orbiting infrared observatory that will take the place of the Hubble Space Telescope at the end of this decade.

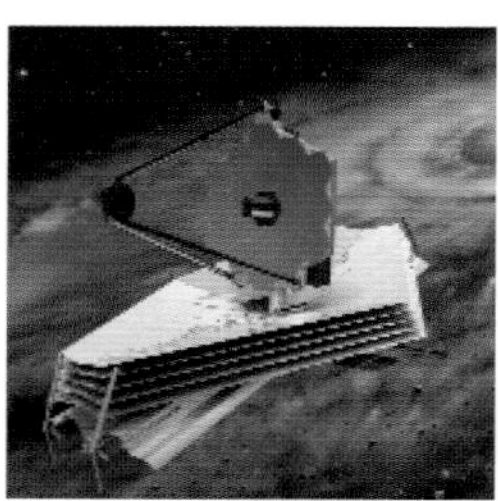

Eric Sawyer explaining the importance of Earth Observation and the role of the GERB instrument at the Meteosat2 launch event, 2005.

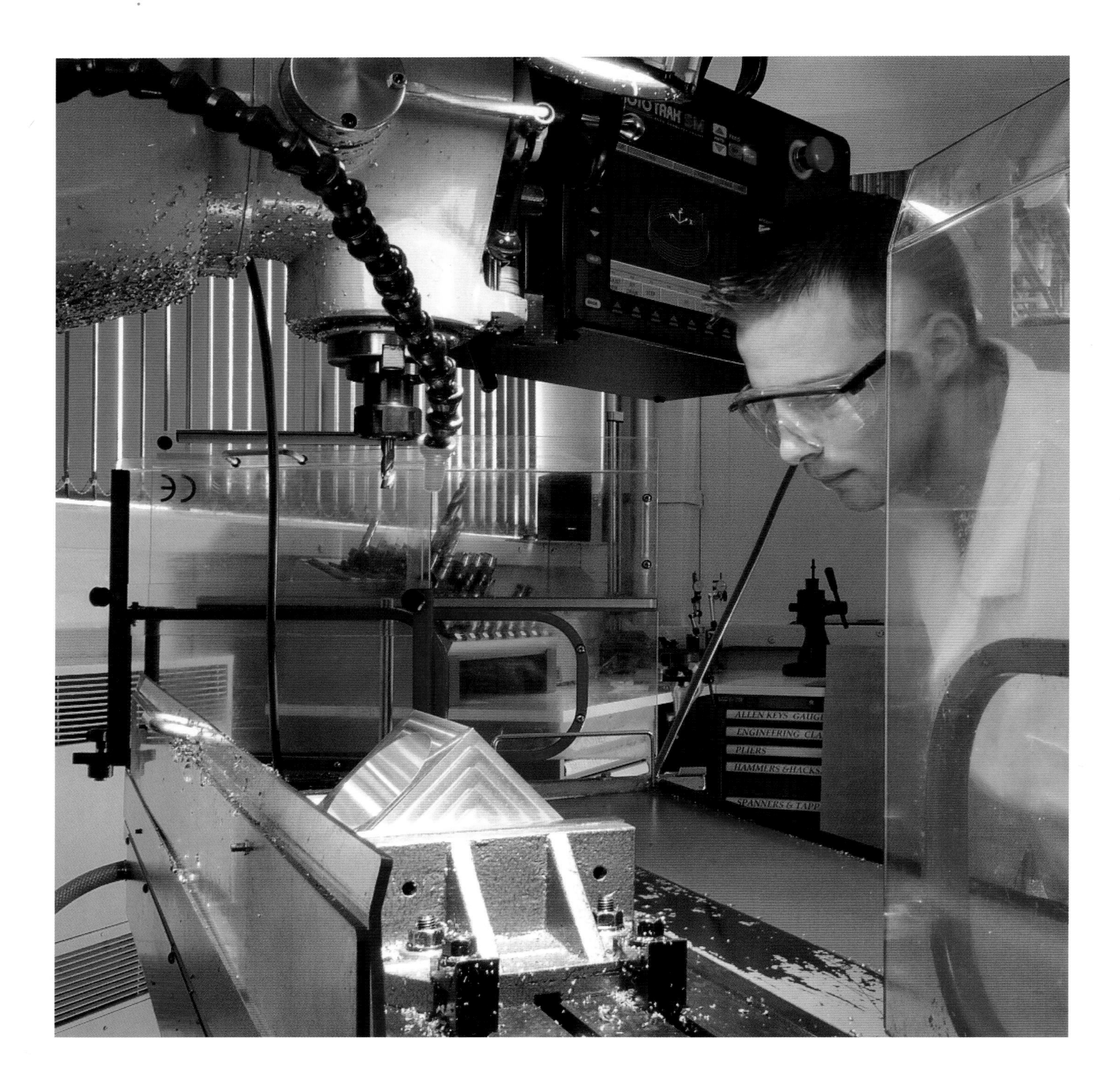
ALLEN KEYS
ENGINEERING
PLIERS
HAMMERS &
SPANNERS &

RapidEye 5-satellite constellation undergoing pre-flight tests, 2007. Built by Surrey Satellite Technology Ltd., RapidEye is a small commercial satellite mission that will enable unprecedented global monitoring of the Earth's surface. Each spacecraft measures less than 1 m^3 and weighs only 150 kg. All five spacecraft will be launched on-board a Dnepr launch vehicle from Baikonur Cosmodrome, Kazakhstan, later this year.

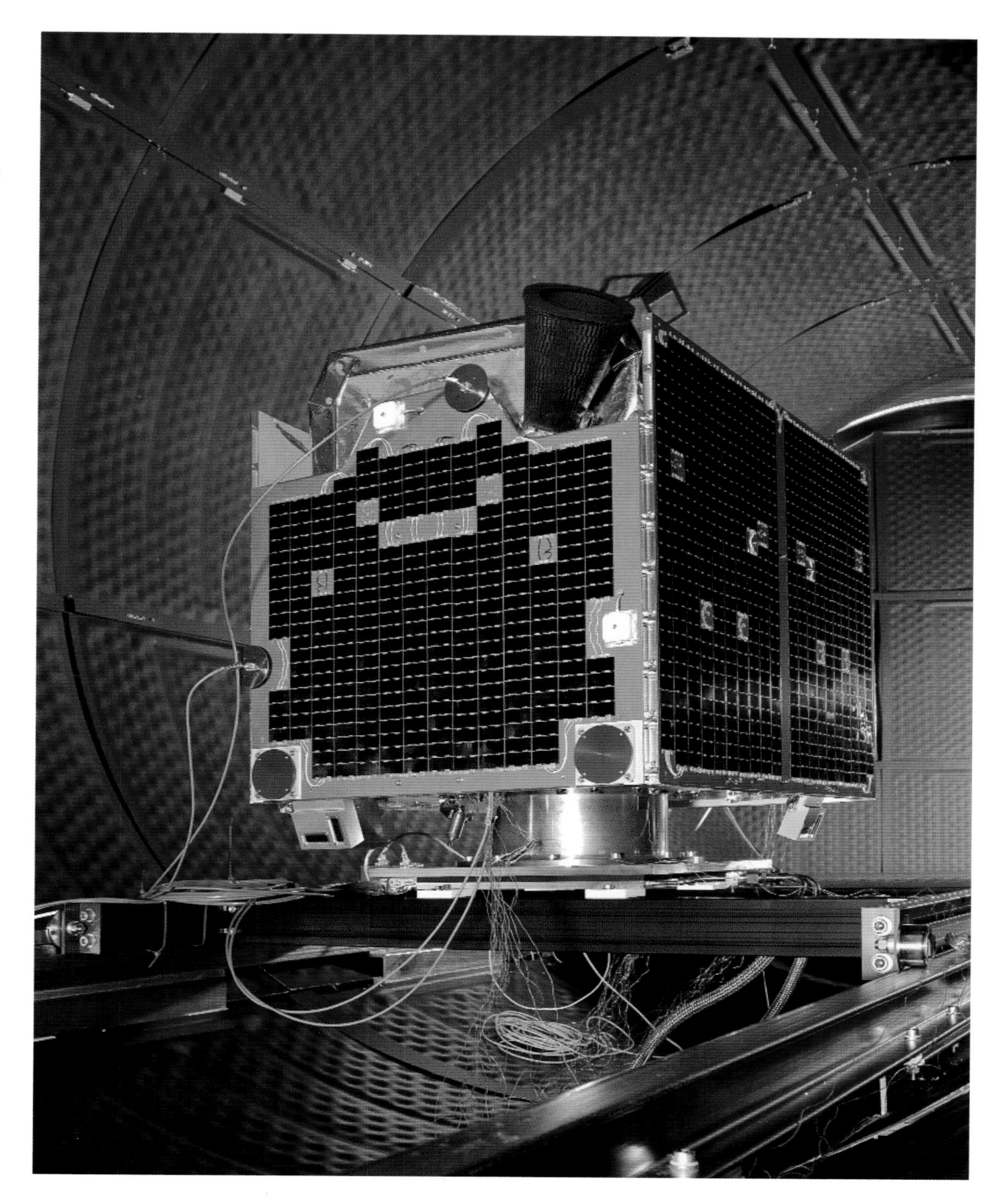

Ben Green, then an apprentice, working in the Project Support Facility, 2006. The facility manufactures, develops, modifies and rectifies flight and non-flight components. Ben is now working in the Cryogenics Group in Technology.

1957

1975

1984

An evolving landscape

1993

2005

2006

Celebrating fifty years of excellence.